WHEN WILL BROCCOLI TASTE LIKE CHOCOLATE?

YOUR questions on **GENETIC TRAITS**

Answered by
Stanford University scientists

Edited by
Dale L Bodian, PhD, and **D Barry Starr, PhD**

Cover Design: Inka Mathew (Green Ink Studio)

Cover Photograph: Dale Bodian

Interior Design: Inka Mathew (Green Ink Studio)

Illustrations: Inka Mathew (Green Ink Studio)

Funded by the Department of Genetics, Stanford University School of Medicine.

ISBN-13: 978-1477578711

ISBN-10: 1477578714

Contact: askageneticist@thetech.org

TABLE OF CONTENTS

Preface

Dale: "Someone was murdered, and it's our job to figure out who did it." As I say these words, I'm kneeling by a body-shaped chalk outline, surrounded by school kids. We're in The Tech Museum of Innovation in San Jose, California, and I'm demonstrating how scientists use DNA evidence to help solve crimes.

Once a week, I and other Stanford University scientists leave our research laboratories to share our enthusiasm for science with school groups and other visitors at The Tech Museum. We run hands-on demonstrations of the experimental techniques we use in our research, and write articles for non-scientists about the latest discoveries in genetics. We also answer questions about genetics submitted to the museum's online exhibit called "Ask a Geneticist."

When it was my turn to answer questions for "Ask a Geneticist" for the first time, I reviewed the questions and answers that had already been written. I was impressed by how creative and insightful the submitted questions are, and how the answers are both interesting and understandable. Although the "Ask a Geneticist" website is well-visited, there are many people who are unaware of this great resource. Collecting some of the best questions and answers into a book, along with updated and new material, seemed like a good way to share them with a wider audience.

Barry: Science is this incredibly exciting, fascinating exploration of how the world works. It should be as popular as any game or crime-solving show. But it isn't.

In school, all the joy is often sucked out of science. It becomes a litany of unpronounceable words that you have to learn the meaning of and dead white guys you have to learn about. It is astonishing anyone enters science!

These students then go on to graduate and avoid science and science topics like the plague. Or they misunderstand science and latch onto questionable theories that can have real consequences for society.

Genetics is a case in point. Everyone should be interested in genetics because a huge part of who they are can be found in their genes and DNA. And yet, school has so failed us that when we did a focus group with 7th graders and asked them what they thought of when they heard the word genetics, one of the most common answers was that it was abstract and had nothing to do with them.

Back in 2003 I set out to try to change this. I wanted to get people more interested in science and so chose to focus on genetics. As I said, genetics seems like a natural entry point into science because it is incredibly relevant to each of us.

Unfortunately, genetics is filled with jargon and confusing topics. Heck, I know graduate students who struggle with two-gene Punnett squares! What hope does anyone else have?

I decided to attack this problem on two fronts. First I would create a set of activities and a website that shows how relevant, interesting and yes, exciting, science can be. Each year we enlighten tens of thousands of visitors at the museum and over a million unique visitors at our website.

Second, I would create a training program where young scientists can be taught how to communicate science to the public in a way that the public will understand and want to hear. The hope is that these scientists will go on, Johnny Appleseed-like, and create similar programs around the U.S. and the world. This is how the Stanford at The Tech program was created.

Graduate students from Stanford University guide museum visitors through hands-on activities at the museum. Through this exercise, they learn what vocabulary they can use, how to target a discussion to a particular audience, the importance of analogies, and so on.

They also receive training in how to write science for the public. The answers in this book sprang from this learning experience.

One of the most popular sections of the "Understanding Genetics" website is "Ask a Geneticist." People at The Tech Museum and all over the world send us questions and we answer them. Some questions are given to Stanford graduate students and staff who write the answers and post them online.

Our goal is to write the answers in a breezy, entertaining way that seeks to demystify science and show how fun it really is. And since the topics come from the public, we know that at least someone is actually interested in them.

Hopefully through a cadre of communications-trained scientists and interesting, easy-to-read content, people will come to realize how fun and exciting science is. Science rocks and everyone should know.

Acknowledgements

We would like to thank Abbie Friedrichs, Carol Bodian, and Mary Tebo for comments on the manuscript, and Debbie Katz and Hank Greely for helpful discussions. We also appreciate the assistance provided by Doug Erickson, Jerry Bodian, John and Nancy Etchemendy, Mark Friedrichs, Mike Cherry, Mimi Calter, Paul Mitiguy, Rick Myers, and Toby Freedman. This book would not have been possible without the visitors to The Tech Museum and to the "Understanding Genetics" website who submitted questions to "Ask a Geneticist," and the scientists participating in the Stanford at The Tech program, who wrote the answers. We also thank the faculty of the Department of Genetics, Stanford University School of Medicine, for their commitment to public science education. A very special thank you goes to the department chairman, Dr. Michael Snyder, for his unwavering enthusiasm and support for this project. Stanford at The Tech is funded by the Department of Genetics, with initial funding provided by a Science Education Partnership Award (SEPA) from the National Institutes of Health (NIH).

Introduction

All science is about aspects of the world we live in and genetics in particular touches our lives in personal ways. Why do I have freckles? How can two people with dwarfism have a child that is of average height? How did some people get extra fingers or toes? From the everyday to the unusual, genetics is at the heart of much of what we see around us.

Genetics begins with DNA. DNA is the sequence of chemical letters that contains the instructions for making you, me, and all living things. Around 99.5% of the 3 billion letters of DNA in a person are identical in everyone. A big part of what makes you different from me can be found in the 0.5% difference. Some of these differences are in genes, the parts of DNA with instructions for a specific trait. Other differences are found in between genes and signal when to turn on certain genes and how much they should be turned up.

Scientists first figured out all of the DNA (the genome) from a person in 2000. This information has led to many unexpected findings about our genes and how traits are passed on from parents to their children. As usually happens in science, finding answers to some questions leads to even more questions.

We are learning that DNA is not a static carrier of genetic information. Through something called epigenetics, DNA is chemically modified so that the same genes can be used differently to give different results. And we are finding out that changes in DNA between generations or even in the same person are much more common than we thought.

All of this means that what we learned in high school or even college biology is not the whole story. We still inherit our genes from our parents— half come from mom and half from dad. But chemical changes in the DNA, interactions between genes, and influences from the environment can lead to surprising outcomes. For example, unlike what you may have learned in school, sometimes two parents with blue eyes can have a brown-eyed child.

These discoveries are ushering in a new era of scientific innovation. Rice has been created that will prevent hundreds of thousands of cases of blindness each year. Patients are being given the right medicine in the right amounts based on their particular set of genes. Crimes are being solved and innocent people set free. All because of our deeper understanding of genetics.

Of course, knowledge is always a two-edged sword. Accompanying the advances in genetics is a host of legal, political, social, and medical issues. Should police be allowed to take DNA samples without consent from anyone who is arrested, whether or not they are convicted of a crime? Should people be tested for diseases for which no treatment is currently available? Should genetically modified food be regulated differently than traditionally bred food? For society to resolve these issues in an intelligent way, non-scientists need to understand recent scientific discoveries and their implications. And also their limitations.

These needs, combined with a desire to share our excitement about genetics with non-scientists, motivated the "Genetics: Technology with a Twist" exhibition at The Tech Museum of Innovation in San Jose, California. A key part of the accompanying website, "Understanding Genetics," is "Ask a Geneticist." There, museum visitors and web users from all over the world ask genetics questions that are then answered by Stanford University scientists. The "Ask a Geneticist" website currently has over 400 questions and answers and receives approximately 1.4 million visitors per year.

This book is a collection of 21 of the best questions and answers from the "Ask a Geneticist" exhibit. The questions were asked by people ranging in age from middle school students to adults. They are about things we encounter in everyday life, like eye color, freckles, the television show CSI, and even poop. There are also more unusual topics, such as two-headed snakes and superheroes.

The answers to the questions are intended to be understood by people without scientific training. The book is not filled with scientific jargon but the answers do use correct scientific terms for things; scientific words are defined in the answers and in the glossary. The science is up-to-date and accurate to the best of our knowledge.

This book is for people who are curious about genetics and want to have fun learning about it. All of the question and answer pairs are independent. They can be read in any order, so feel free to focus on those topics of most interest to you.

Want to find out more? Or have a question of your own to ask? Visit our website! **http://genetics.thetech.org/ask-a-geneticist**

SNIPS AND SNAILS AND PUPPY-DOG TALES

Questions about plants, animals, and unusual creatures

> *I read that many bony fish including Nemo, the clownfish, change sex. How does changing sex impact their DNA? Were they born with both X and Y chromosomes or were they somehow changed when the sex was changed?*
>
> A curious adult from California

IT IS AMAZING the variety of ways Mother Nature has come up with to determine whether an animal is going to have a boy or a girl. As you know, for humans, sex is determined by the presence of a Y chromosome—humans with an X and a Y chromosome are usually male and those with two X chromosomes are usually female.

In birds, the opposite is generally true. The male has two of the same chromosome, Z, while the female has two different sex chromosomes, W and Z. So in birds, the female decides the sex of the offspring while in humans, it is the male.

Other animals have no sex chromosomes and their sex is determined in different ways. For example, the temperature at which their eggs are incubated determines the sex of a turtle—if the temperature is below 86 °F, they are all males, above 86, and they are all females.

For clownfish like Nemo, it is particularly complicated. All clownfish are born male. A clownfish group consists of a dominant male and female and 0–4 juvenile males. So where did the female in the group come from? When the female dies, the dominant male becomes the dominant female and one of the juveniles becomes the dominant male.

Do these ways of making boys and girls have anything in common? Yes. In all cases, whether you are a boy or a girl is determined by a certain set of genes being turned on.

Genes determine sex

In people, certain genes on the Y chromosome determine this. For example, the Y chromosome has a gene called *SRY* that signals the body to become male. In other words, the *SRY* gene must be turned on to make a male. In fact, if *SRY* is present in someone with two X chromosomes, they appear male and if someone is XY but has a mutated *SRY* gene, they look female.

The same is true for species without sex chromosomes. For example, in turtles it may be that high temperature shuts off the turtle *sry* gene so you get females. In the case of clownfish, the absence of a female results in a male changing to a female. Perhaps a female clownfish has some sort of chemical that keeps males from becoming female.

In terms of what's going on biologically in the clownfish, apparently the dominant male has functioning testes and some latent cells that can become ovaries under the right conditions. Once the female dies, the testes in the dominant male degenerate and ovaries form from the latent ovarian cells. Voilà, he is now she.

In terms of the actual biological sculpturing involved, not much is known yet. I hope this helped shed some light on the topic—it is a strange, strange world out there.

Answered by Dr. Barry Starr, Stanford University

> *Why can't mules breed? I understand that a horse and a donkey make a mule but why can't two mules have a baby mule?*
>
> A middle school student from Michigan

YOU'RE RIGHT, a horse and a donkey can have kids. A male horse and a female donkey have a hinny. A female horse and a male donkey have a mule.

But hinnies and mules can't have babies of their own. They are sterile because they can't make sperm or eggs.

They have trouble making sperm or eggs because their chromosomes don't match up well. And, to a lesser extent, because of their number of chromosomes.

A mule gets 32 horse chromosomes from mom and 31 donkey chromosomes from dad for a total of 63 chromosomes. (A horse has 64 chromosomes and a donkey has 62.)

To understand why this is a problem, we need to understand how sperm and eggs are made. And to understand that, we need to go into a bit more detail about chromosomes.

Remember, we have two copies of each of our chromosomes—one copy from mom and one from dad. This means we have two copies of chromosome 1, two copies of chromosome 2, etc. However, this isn't entirely true for the mules.

The mule has a set of horse chromosomes from its mom. And a set of donkey ones from its dad.

The mule's chromosomes aren't really matched sets as in a horse, a donkey, or a person. In these cases, a chromosome 1 is very similar to another chromosome 1. It looks pretty much the same and has nearly the same sequence of A's, G's, T's, and C's. For example, two human chromosome 1's differ only every 1000 letters or so.

But a donkey chromosome doesn't necessarily look like a horse one. And the poor mule even has an unmatched horse chromosome just sitting there.

Meiosis

To make a sperm or an egg, cells need to do something called meiosis. The idea behind meiosis is to get one copy of each chromosome into the sperm or egg.

For example, let's focus on chromosome 1. As I said, we have one from mom and one from dad. At the end of meiosis, the sperm or egg has just a single chromosome 1. Not two.

This process requires two things. First, the chromosomes have to look pretty similar, meaning they are about the same size and have the same information. This is important for how well they match up during meiosis.

And second, at another critical stage of meiosis, there have to be four of each chromosome. Neither of these can happen completely with a mule.

Let's take a closer look at meiosis to see why this is. The first step in meiosis is that all of the chromosomes make copies of themselves. No problem here—a mule cell can pull this off just fine.

So now we have a cell with 63 doubled chromosomes. It is the next step that causes the real problem.

In the next step, all the same chromosomes need to match up in a very particular way. So, the four chromosome 1's (mom's chromosome 1 and its double, plus dad's chromosome 1 and its double) all need to line up together. But this can't happen in a mule very well.

As I said, a donkey and a horse chromosome aren't necessarily similar enough to match up. Add to this the unmatched chromosome and you have a real problem. The chromosomes can't find their partners and this causes the sperm and eggs not to get made.

So this is a big reason for a mule being sterile. But how is the silly thing alive at all?

Well, there are a couple of reasons. First, having an odd number of chromosomes doesn't matter for everyday life. A mule's cells can divide and make new cells just fine. Which is important considering a mule went from 1 cell to trillions of them!

Chromosomes sort differently in regular cells than they do in sperm and eggs. Regular cells (called somatic cells) use a process called mitosis.

Mitosis

Mitosis is like the first step of meiosis. The chromosomes all make copies of themselves. But instead of matching up, they just sort into two new cells. So for the mule, each cell ends up with 63 chromosomes. No matching needs to happen. And our lone horse chromosome is fine.

The other reason a mule is alive is that nothing on the extra or missing chromosome causes it any harm. This seems obvious at first except that usually having extra DNA causes severe problems. In people, extra chromosomes usually result in miscarriages. Sometimes though, a child can survive with an extra chromosome.

For example, people with an extra chromosome 21 have Down syndrome. Having extra genes from that extra copy of chromosome 21 causes the symptoms associated with Down syndrome.

So having extra chromosomes often leads to real problems. But the mule is by and large OK.

The extra genes must not be that big a deal for the mule. In other words, the extra genes on the horse chromosome do not cause problems for the everyday life of a mule.

So mules are sterile because horse and donkey chromosomes are just too different. But they are alive because horse and donkey chromosomes are similar enough for the animals to breed.

Answered by Monica Rodriguez, Stanford University

What would it take to develop a superpower?

A middle school student from New York

WHAT A RELEVANT QUESTION in today's world of Superman, Spider-Man, X-Men, and, of course, the TV series, Heroes! Are there really superheroes in this world? Do we have the science to create them?

The quick answer is that we probably can't. We aren't very good at changing our genes in any specific way. And any new power we could create would probably come with so many bad side effects they aren't actually that super anyway. Let's dig a bit deeper and see why this is.

First off, we aren't going to be able to make any of the really fancy superheroes. No one is going to be flying like Superman. Or controlling magnetic waves like Magneto. Or controlling fire like the Human Torch.

This is because these sorts of things are way too complicated for a single gene to handle. Any one of these new skills would need lots of changes in lots of genes (if they were even possible). To understand why, think about something simple like walking.

Walking involves a signal from our brain that travels through our back (spinal cord) and orders our leg muscles to move. Every part of this circuit involves many genes working together to make us take a step forward.

Now imagine what it would take to make someone fly. We would have to change genes in our brain, our nerves, our back, our bones and our muscles. That is a lot of change! Too many to have any chance of happening all at once either in nature or in the lab.

Our superhero goals have to be more modest. Imagine our superheroine, Lady Justice.

With just three DNA changes we can make her strong, super-brilliant, and feel no pain. What a superheroine!

Or is she? Let's look at each new skill in more detail. I think you'll see that these skills have worse tradeoffs than having a secret identity. Or anger management problems.

Ouch! genes

First the fact that Lady Justice feels no pain. What a great skill this would be. She can fight all day long and not stop because it hurts. And no amount of torture will make her give up the location of the secret hideout.

But feeling no pain is not as great as it sounds. There are people in a family in Pakistan who are just like this. Scientists worked with this family and discovered that they have a mutation in a gene called SCN9A. This gene normally makes an ion channel, which is like a gate that lets sodium ions enter a nerve.

Sodium ions are really important to tell nerves about the outside world. But, because of this mutation, the gates on these nerves don't work. So, these kids never feel the sensation of pain.

Like I said, how awesome for a superhero. Or would it be?

Not really. Although these kids don't feel pain, they still get hurt and injured. The bad part is that they don't know they are hurt or injured (because they don't feel pain) and don't seek medical help. So their injury keeps getting worse and worse. So, Lady Justice would probably be better off feeling at least some pain.

Smart genes

Well, OK, that one isn't great. But super intelligence has got to be useful, right? All the great hero teams have at least one super smart guy.

And there may be DNA changes that can cause this to happen. Scientists at Princeton University created a special kind of mouse, called a "Doogie" mouse.

The scientists changed a single gene called *Grin2b* in this mouse. The protein encoded by *Grin2b* is, yes, you guessed it, another ion channel!

This time, it is an ion channel that lets calcium ions enter the neuron. Normally, this gene is present in children. This is what makes kids so much smarter than adults and able to pick things up so quickly. By the time we grow up, though, this gene is turned off.

Scientists changed this gene so that it keeps working in adult mice. And, lo and behold, these mice were much smarter than normal adult mice.

Scientists have a way of testing mice, kind of like "mouse SATs" and these smart mice did much better at those tests. They had much better memory and learned faster than their buddies.

Maybe the same sort of thing could be done in people one day. But let's pause for a minute here and see if this is really a good thing to have. Yes, it would be good to be super smart. But, do we really want to remember everything that happened in our lives?

Probably not. Again, what seems like a good thing on the surface is not a good thing in the long run.

Genes for strength

Maybe being super strong would be better. Not Superman strong but maybe "The Hulk" strong. Or like a 4-year-old child who can lift 7 pounds with his arms extended.

Whew, that is one strong boy. And you can find him in Germany.

He has a mutation in a muscle gene called myostatin. Because of this mutation, his muscles have grown extraordinarily strong and he is certainly going to be a kid to watch out for.

But, again, you don't get this for free. Being this strong is not great for his heart. Doctors are keeping a close watch on this little tot because heart muscles are very important for our day-to-day function. And improper heart muscles can lead to heart attack.

Let's take a look at our proud superheroine. She is pretty beat up because she feels no pain. And so doesn't know she's hurt.

She is smart but kind of distracted since she remembers everything.

And she has bulging muscles. But has to be careful because she has a bulging heart too. Lady Justice isn't too super is she?

So it is pretty unlikely that we can use genetics to become a superhero. Certainly not with what we know now. The genetic changes we know about that might give a person superhuman powers always come at a cost.

And not all superheroes are accidents of nature or created in a lab. Some people turn themselves into superheroes the old fashioned way—with willpower and hard work.

Have you heard of Grant Hackett? He is a superhero swimmer who has won major world swimming competitions several times.

Grant is a normal guy with extraordinary willpower. He trained really hard and improved his lung capacity so much that he could use his really strong lungs to breathe deeply and swim fast. So, here is one example where a normal athlete did something like a superhero. Not because he had some "cool" gene but because he trained hard. And, there is no bad side to his improved lung capacity.

And to the people who came before us, we're all superheroes. Imagine you are a caveman living many thousands of years ago.

There you are living in a cave. What would you imagine as superhero powers? Probably things like controlling fire, or having as much food as you want. Yup, to Cro-Magnon man, we are the superheroes.

Answered by Dr. Shalu Srinivasan, Stanford University

How are two-headed snakes possible?

A middle school student from California

WOW, THERE AREN'T JUST two-headed snakes, but two-headed sheep, pigs, cats, dogs, fish, and even people! With two separate heads (and two separate brains), how can they survive? Wouldn't the two heads fight with each other? And how would they control their body?

Well, they do have a lot of problems and most of them don't survive very well in the wild. They can't decide which direction to go, and sometimes they even try to attack and swallow the other head!

Two-headedness is really a severe example of conjoined twins. These twins happen when identical twins don't separate completely. Instead of two twins, you end up with twins who are still attached.

The attachment can be as little joining as just a piece of skin or cartilage connecting the two twins. Or in more severe cases, they can share a whole body but separate heads!

Ending up with two heads

Identical twins happen when a single egg is fertilized but then splits into two separate embryos. Each embryo then goes on to grow into a separate person. Depending on when the split happens, twins can either grow in their own sac with their own placenta or they can share a sac and placenta.

Conjoined twins can happen in this second case. But it isn't common...in humans it only happens in around 1 in 50,000 to 200,000 births.

There are two theories for how conjoined twins come about. One is that they happen when the fertilized egg splits incompletely. Another is that the embryo split is complete but for some reason the two new embryos re-fuse later on.

Twins can end up fused at various places. Where the twins end up attached depends on a lot of factors, like how they were oriented in the womb, the time and location of splitting, etc.

Having conjoined twins doesn't usually run in families. This is because identical twinning doesn't run in families either. It is a random event.

But some cases of having extra limbs can be genetic. One of the best understood is polydactyly, which is when people have extra fingers and toes.

Genetic causes of duplication

There are lots of genetic ways to end up with extra fingers and toes. Here I'll focus on just one case.

Most cases of polydactyly can be accounted for by changes in just one of a few different genes. This is surprising because making fingers is such a complicated task. Lots of different genes need to work together to make ten fingers and ten toes.

So how can just one change in a gene cause something like an extra finger? Because this one gene can affect lots of other genes, too.

Recall that genes are our basic instruction manual for building and running all the things in our body. These genes are "read" by cells and made into proteins. Each protein then goes on to do a specific job in the cell.

Some proteins like hemoglobin carry our oxygen. And others like insulin help us use the food we eat.

It's very important that the right proteins get made and in the right amount. If we don't have enough hemoglobin, we have trouble breathing. Or if we don't have enough insulin, we get the disease diabetes.

That's why cells have lots of ways to control when, where, and how often a gene is read. One of the ways cells do this is by using proteins called transcription factors* (TFs).

TFs stick to a gene and control how often the gene is read. Lots of copies of each TF are made and they usually stick to and control a bunch of genes all at once.

Some TFs called activators make cells make extra proteins. And some TFs called repressors cause cells to make less protein.

Well, one reason polydactyly happens is because of changes in a single gene, *GLI3*. The GLI3 protein is one of these repressor TFs.

One of its jobs is to keep two genes, *HAND2* and sonic hedgehog *(SHH)* from being read too often. When the GLI3 protein isn't doing its job right, too much HAND2 and SHH proteins get made. And too many fingers and/or toes get made too.

Some people with polydactyly have mutations in the *GLI3* gene that cause the GLI3 protein either to not work at all or that cause it to work only very weakly. So basically, the embryo ends up with too much SHH and HAND2, causing the extra fingers and toes.

This kind of polydactyly can be inherited from your parents. It is an autosomal dominant trait.

---

* These proteins are called transcription factors because the act of reading a gene is called transcription.

Autosomal means that boys and girls are just as likely to get it (it isn't sex-linked). Dominant means that you only need to have one copy of the polydactyly version to end up with extra fingers and/or toes. In fact, if you have one copy, you will have extra digits!

So if just one parent has this mutation in GLI3, their kids have a 50% chance of getting it too. If both parents have it, their kids have a 75% chance of having extra fingers or toes!

Polycephaly (having multiple heads or faces), or polymelia (having extra limbs) can also have some genetic causes. But, these are usually associated with other serious genetic defects (it's not well understood for either of these cases), or more frequently, are the result of twinning or another problem during development.

Answered by Julia Oh, Stanford University

> **When will broccoli be modified to taste like chocolate, have the antioxidants of cranberries, and be available at my grocery store?**
>
> A curious adult from California

I BET THIS IS the kind of question lots of people have asked. But don't hold your breath because it won't be happening anytime soon.

Why not? First off, we don't really understand how cranberries make their antioxidants or how chocolate gets its flavor.

Second, once we figure out how these things work, it won't be easy to add these traits to broccoli. This is because lots of different genes are involved in making cranberry antioxidants and chocolate flavor.

Right now it is hard to move a single gene from one plant to another. So moving lots of genes would be even harder.

But great science often starts with questions that are seemingly impossible to answer. So let's take a stab at it.

Antioxidants vs. free radicals

There are lots of things around us that can hurt our bodies. One of the real troublemakers is free radicals.

In our bodies, free radicals are molecules that really like to react with other molecules that they are not really supposed to react with such as our very own DNA. When they do so, they damage DNA, which is not good for our cells and our bodies. This causes lots of problems including cancer and aging.

Free radicals come from lots of different things including the food we eat.

Luckily for us, nature has given us ways to fight these free radicals. They are called antioxidants. Antioxidants help keep free radicals away from our DNA and cells by reacting with them. They also help make the free radicals less dangerous.

Studies show that antioxidants can help prevent cancer and heart failure. They might even be able to slow down aging! So, they are pretty good things to have.

So where do we get them? Well, our bodies make some but we get a lot of them from some foods we eat. As you've noted, fruits like cranberries have lots of antioxidants in them. Both our bodies and foods make antioxidants the same way—with genes.

Genes and antioxidants

Genes are really just instructions for making proteins. And proteins are the things that do most of the work in our cells (and bodies) including getting rid of free radicals.

There are several kinds of antioxidants and genes play an important role in making all of them. Some genes have the instructions for directly making proteins that get rid of free radicals. An example is superoxide dismutase.

Small molecules that are not proteins are another class of antioxidants. It usually takes many proteins to make these molecules. Which means it takes many different genes too. Vitamin C is one of these antioxidants.

Our foods, like us, have both kinds of antioxidants. But the most useful ones from foods are the small molecules. Our stomachs digest the proteins that directly eliminate free radicals. But the small antioxidants can survive.

Unfortunately, the small molecules are the trickiest to get plants to make. Why? Because it takes more than one gene to make this kind of antioxidant. And the more genes you need to put into a plant, the harder it is to do.

This doesn't mean it is impossible. For example, scientists have been able to make a new kind of rice that makes tons of vitamin A. This new kind of rice helps prevent blindness. It is called "Golden Rice" because of its yellowish color.

How did scientists make this rice? They took two genes that help make vitamin A and put them into the rice plant. This "simple" task took over 7 years of work and $2.6 million. So while not impossible, it isn't easy either. Or cheap!

We can't really replicate this with other antioxidants right now because scientists haven't yet worked out the genes plants use to make them. Once they do, then scientists might be able to make a broccoli with more or additional antioxidants. As long as there is a market for it that is.

Chocolate broccoli

This story now brings us to chocolate-flavored broccoli. No doubt there would be a market for that! But this one is even less likely than a broccoli with the antioxidants of cranberries.

Why? Mainly because we don't yet really understand why chocolate tastes the way it does.

Part of the reason for chocolate's bitter taste seems to be a group of molecules known as flavanols. These chemicals are also antioxidants and are found in green tea, nuts, red wine, and other foods too.

Of course, not all of these molecules taste the same. So scientists need to find which kinds of flavanols chocolate has.

Then, scientists will have to find the genes that make the flavanols of chocolate and then they will need to put them into the broccoli. What this means is that it will definitely be a while until there is chocolate broccoli.

But all is not lost! Scientists are sequencing the genome of the cacao plant (from which cocoa, and hence chocolate, is made). In other words, scientists are working on the hard job of reading the entire DNA of the cacao plant.

Hurray! They are three years ahead of schedule and have already made part of the genome publicly available. Now that scientists know more about the cacao plant's genome, they may be able to figure out how it makes its chocolate flavor.

Of course, even if we got chocolate's flavor into broccoli, it still wouldn't be a chocolate bar. No one just eats plain cocoa beans.

Chocolate has many ingredients. To make chocolate the cocoa beans are first processed. Then, milk, sugar, and other things are added. Imagine a broccoli that also makes milk! Not likely. If we do all that to broccoli, it suddenly isn't so healthy either.

After going through all of this, I think you can see that it will be quite a chore to make broccoli like you want it! Most likely it probably won't be worth a scientist's effort (at least not in the near future).

Of course, there is another option. You could take some broccoli, a few cranberries, a bar of dark chocolate, and add it all into a blender. This wouldn't be too tasty but it would have all the ingredients of the broccoli you're looking for. Me, I'll stick with cranberry scones for breakfast, broccoli at dinner time, and dark chocolate for dessert!

Answered by Jose Rafael Morillo Prado, Stanford University

DON'T IT MAKE MY BLUE EYES BROWN

Questions about eye color

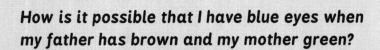

How is it possible that I have blue eyes when my father has brown and my mother green?

A high school student from Illinois

IF THERE WERE just one eye-color gene, then your situation would be pretty uncommon. And yet your situation can and does happen. A lot.

A brown-eyed dad and a green-eyed mom can have a blue-eyed child because there are at least two eye-color genes. Because of this, it is possible for both green- and brown-eyed parents to be carriers for blue eyes.

And as carriers, they each can pass down blue-eye genes to their children. Let's go into a bit more detail about how all of this works.

Eye-color genes

As I said, eye color happens because of at least two separate genes. The first gene, *OCA2*, comes in two versions, brown *(B)* and blue *(b)*. The second gene, called *EYCL1,* also comes in two versions, green *(G)* and blue *(b).**

Here is a table that shows how these genes are thought to work together to create eye colors. The first pair of letters in every combination represents versions of *OCA2* (the brown-blue gene) and the second pair represents versions of *EYCL1* (the green-blue gene).

---

* This is a simplified version of how things actually work. There are undoubtedly more genes that we haven't identified that are also involved in eye color. But these two genes are enough to answer this particular question. See the answer starting on page 26 for a more complex view of eye color.

GENE VERSIONS	EYE COLOR
BB bb	Brown
BB Gb	Brown
BB GG	Brown
Bb bb	Brown
Bb Gb	Brown
Bb GG	Brown
bb GG	Green
bb Gb	Green
bb bb	Blue

The first thing to notice from this table is that whenever there is a *B*, there are brown eyes. So *B* is dominant over both *G* and *b*. Also, whenever there is a *G* (but no *B*) there are green eyes. So *G* is dominant over *b*.

A couple of things might seem weird here. First, there are two separate genes and yet *B* from one gene is dominant over *G* from another gene.

The other odd thing is that the recessive forms of both genes are blue. These two issues are related.

Eye color comes from melanin

Eye color happens because of the amount of the pigment melanin found in the eye. Not anywhere in the eye but in a very special place, the stroma of the iris.

Lots of melanin in the stroma of the iris gives brown eyes and less melanin gives green. Little or no melanin there gives blue eyes.

So that is why brown is dominant over green. The *B* version of *OCA2* tells the eye to make lots of melanin. The *G* version of the *EYCL1* gene tells the eye to make some. What happens if both are present? Lots of melanin gets made which means brown eyes.

The fact that both recessive forms are blue makes sense from this as well. The recessive forms of these two genes are recessive because they don't work. A broken *OCA2* gene is like a broken *EYCL1* gene—no melanin gets made in the stroma. No melanin in the stroma means blue eyes.

OK, so now we see why brown is dominant over green. And why blue is recessive to both. But we still haven't explained your situation.

Mixing gene versions

We can figure out your situation by noticing something else in the table— all the eye colors have two versions of each gene. There are two copies of *OCA2* and two copies of *EYCL1* in each case.

This is because we have two copies of most of our genes, one from mom and one from dad. It is this fact that allows for a brown-eyed dad and a green-eyed mom to have a blue-eyed child.

Let's look at *OCA2* as an example. If someone has two *B* versions, then they have brown eyes. And if they have two *b* versions, then they don't have brown eyes (they'll have either green or blue). But what if they have one *B* and one *b*?

Then they usually have brown eyes. But half of the time they will pass the blue version to their kids. And if the other parent passes the *b* version of *OCA2* as well, then the child will not have brown eyes. This all works for the *EYCL1* gene too.

So in your case, the easiest way to explain your blue eyes is if both your mom and dad are carriers for blue eyes. Your mom is most likely *bb Gb* and your dad is either *Bb Gb* or *Bb bb* (we can't tell the difference).

Since you have blue eyes, each parent gave you a *b* from *OCA2* and a *b* from *EYCL1*. The end result is that you are *bb bb,* which is blue eyes.

So there you have it. Now you know how it is possible for a brown-eyed dad and a green-eyed mom to have a blue-eyed child. We'll learn more about how eye color works in the next chapter, and see that it involves even more than two genes!

Answered by Dr. Barry Starr, Stanford University

> *Is it possible for two blue-eyed parents to have a green- or brown-eyed child? I thought I knew enough genetics to know that blue + blue yields ONLY blue eyes.*
>
> Curious adults from around the world

YES, BLUE-EYED PARENTS can definitely have a child with brown eyes. Or green or hazel eyes for that matter. If you stayed awake during high school biology, you might find this answer surprising. We were all taught that parents with blue eyes have kids with blue eyes. Every time.

This has to do with the fact that blue eyes are supposed to be recessive to brown eyes. This means that if a parent has a brown-eye gene, they'll have brown eyes. Which makes it impossible for a blue-eyed parent to have a brown-eyed child—they don't have a brown-eye gene to pass on!

And you thought a two-gene model was complicated...

Eye color used to be taught as a fairly simple trait. The idea was that we had one eye-color gene and that this gene came in two varieties—brown and blue. Geneticists represented the brown version as *B* and the blue version as *b*.

This model also said that the blue *(b)* version of the gene was recessive to the brown *(B)* version. This matters because it is an explanation for how brown-eyed parents can have a blue-eyed child.

See, we have two copies of each of our genes—one from mom and one from dad. This means we all have three possible combinations for this eye-color gene: *BB*, *Bb*, and *bb*.

BB is of course brown and in this model, *bb* would be blue. Since blue is recessive to brown, *Bb* people do not have blue eyes. But they can pass a *b* down to their kids so that these kids might end up with blue eyes.

Now eye color is obviously more complicated than this. This model doesn't explain green eyes for example. Scientists added a second gene to try to explain green eyes but we don't need to go into that here (you can read about the two-gene model in the answers starting on pages 22 and 90).

All we need to know is that with this expanded model, if you have a *B*, you have brown eyes no matter what this green-eye gene says. So if this were the case, then we'd expect the following possibilities:

GENE VERSIONS	WHAT IT MEANS
BB	Brown eyes
Bb	Brown eyes
bb	Not brown eyes

Again, *bb* people should not be able to pass on brown eyes to their kids. But we know they can. Which means that this model is incomplete (or wrong).

So something else must be going on. That something is most likely other genes involved in eye color that we don't know about.

Real people are more complex

The results I just put into the previous table are theoretical and based on the one-gene model I talked about. Here are some actual results from a study on Europeans' eye color adapted from 23andMe's website:

GENE VERSIONS	WHAT IT MEANS IN EUROPEANS
BB	85% chance of **brown** eyes 14% chance of green eyes 1% chance of blue eyes
Bb	56% chance of **brown** eyes 37% chance of green eyes 7% chance of blue eyes
bb	1% chance of **brown** eyes 27% chance of green eyes 72% chance of blue eyes

As you can see, the original model holds up pretty well for *BB* and *bb* people. Most *BB* people have brown eyes and most *bb* people don't. But the model clearly doesn't explain the following:

1. 1% of *bb* people have brown eyes
2. 1% of *BB* people have blue eyes (and 14% have green)
3. 44% of *Bb* people do not have brown eyes

The biggest disconnect is with *Bb* people. Only 56% have brown eyes. Part of the reason is that eye color is not the simple decision between the brown (or green) and blue versions of a single gene. There are many genes involved and eye color ranges from brown to hazel to green to blue to... But to simplify the story we'll focus on brown eyes and blue eyes.

Where does eye color come from?

Eye color happens because of the amount of the pigment melanin in the iris of your eye. If you have no melanin in the front part of your iris, you have blue eyes. Lots of melanin results in brown eyes, and less melanin gives green.

As we saw, what we are taught in high school biology is generally true: brown-eye genes are usually dominant over green-eye genes which are both usually dominant over blue-eye genes. However, because many genes are required to make each of the yellow and black pigments, there is a way— called genetic complementation—to get brown or green eyes from blue-eyed parents.

Genetic complementation

The best way to illustrate how this might happen is with an example. Let's say there is a genetic pathway made up of four genes (cleverly named A, B, C, and D) that are needed to make brown eyes. A mutation in both copies of any one of these genes results in blue eyes.

As you may already know, we have two copies of each gene, one from our mom and one from our dad. So we have two copies of the A gene, two copies of the B gene, and so on. If either parent gives you a brown version of a gene, it will usually be dominant over the blue copy.

To make things simpler, I'll write the brown version of a gene with a capital letter and the blue version with a lowercase letter. So *"A"* means the brown version of the A gene and *"a"* means the blue version of the same gene.

Now let's say that dad has blue eyes because he has two *"a"* versions of the A gene. His B, C, and D genes could give him brown eyes but don't because he doesn't have a non-mutated (brown) version of gene A for them to work with. He is *aa BB CC DD*.

Now let's say that mom has blue eyes because she has two *"d"* versions of her D gene. Her A, B, and C genes are versions that could give brown eyes if she had a working copy of gene D. She is *AA BB CC dd*.

Suppose that mom gives you a brown copy of gene A and dad gives you a brown copy of gene D. In other words, you end up as *Aa BB CC Dd*.

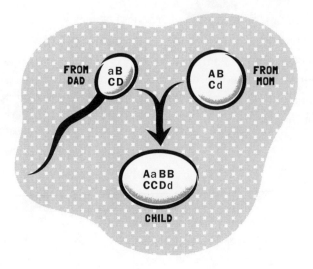

Genetic complementation. A, B, C, and D represent the genes needed to make brown eyes. Between the two parents, the child receives at least one working copy of all four genes and so has brown eyes. *a* and *d* are the forms of genes A and D with mutations that lead to blue eyes when present in two copies.

What color eyes would you have? Brown. Even though both of your parents have blue eyes. Because unlike your parents you have brown-eye versions of all four genes. (The same argument works for green eyes as well.)

Recombination

Another common genetic process that could be responsible for brown eyes from blue-eyed parents is called recombination. This explanation doesn't require any new genes be introduced. It works even with a single gene model!

When eggs and sperm are made, only one of a pair of chromosomes gets put into an egg or sperm. Before this happens, there is a bunch of DNA swapping that goes on between the pair of chromosomes. Sometimes when the DNA is swapped or recombined, DNA mutations get fixed.

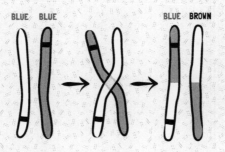

Recombination. Two copies of the eye-color gene each have a different mutation (■) that gives blue eyes. After recombination, one copy has both mutations so is a blue-eye version, and the other copy has no mutations so is a brown-eye version. One blue-eye version and one brown-eye version gives brown eyes.

Again, an example can show how this might work. Imagine dad has blue eyes because of one mutation in each copy of an eye-color gene. But that the mutations in the two copies are in different places in the gene—one copy of the gene has a mutation at its front end and the other copy of the gene has a different mutation at its back end. Each gene copy has one mutation but at different places in the gene.

Now imagine that when dad's sperm is being made, the front part of the eye-color gene is switched between the two copies of the gene resulting in one brown-eye version with no mutations and one blue-eye version with two mutations. Now dad can produce a brown-eyed child. (Again, the same argument works for a green-eye gene as well.)

Environment

A third way to get brown eyes from blue-eyed parents is for something in the environment to affect the eye-color gene or how it works. Even though there are well-documented cases in which this happens, the reasons for it are pretty poorly understood.

There are cases, for example, of certain drugs changing a person's eye color—the environment clearly has changed what happened to eye color in this case. Another possibility is that a gene is on or off for some reversible reason instead of an irreversible change in DNA. In this case, something in the environment reverses the change, turning the eye-color gene back off or on.

Sometimes this is due to chemical changes in DNA that don't change the sequence. This is called epigenetics. "Epigenetics" means "beyond genetics" and can explain many environmentally induced changes in gene expression.

Well, I hope this helped answer your question. As you can see it is all pretty complicated. Scientists are looking for new pigmentation genes and hopefully their studies will help us understand how these genes work together to produce all the different eye colors we see.

Answered by Ky Sha and Dr. Barry Starr, Stanford University

Can you explain the genetics of an individual with one blue eye and one brown eye? David Bowie is one of these individuals.

A curious adult from Tennessee

GREAT QUESTION. Different-colored eyes are pretty rare in people although it is more common in some animals. For example, dogs like Siberian Huskies and cats and horses often have different-colored eyes because of inbreeding.

But this sort of thing obviously happens in people too. Some are even famous like David Bowie, Jane Seymour, and Christopher Walken.

Different-colored eyes—or heterochromia—works differently than other eye colors. It tends to happen when something has gone a bit wrong in making eye color.

DRIVER LICENSE
DL 22335678
NAME: SMITH, JOHN DOE
ADDRESS: 123 EASY STREET
TOWNVILLE, CA 94567
DOB: 01/02/2012
ISSUED: 03/04/2012
SEX: M
EYES: BLUE & GREEN

John DoeSmith

Heterochromia can be the result of too much or too little pigment in the iris (or part of the iris) of one eye. This can be because of genetics or because of disease or injury.

See, eye color usually comes from the amount of pigment in the front part of the iris of the eye (called the stroma). Little or no pigment gives blue eyes, some pigment gives green, and lots of pigment gives brown eyes.

The pigment is actually made in special cells called melanocytes. If anything happens that affects the health of the melanocytes in an eye, then that eye will look blue. No melanocytes means little or no pigment which means blue eyes.

Another way to end up with different-colored eyes is if each eye has a different set of eye-color genes. This isn't common but we'll go over it as well.

So how can one eye end up without any melanocytes? There are a number of ways that all have to do with how sensitive a melanocyte is.

Damaged or lost melanocytes

Melanocytes are made in the growing fetus. Once made they need to migrate to the right spots such as eyes and skin. If something affects their growth or their ability to travel, then they die off.

To end up happily making pigment in the right place, melanocytes need a variety of signals from other cells. This network of cells is pretty easily disrupted.

Imagine that a fetus suffers some sort of injury to the left side of the head while in the womb. This injury kills off some key cells that are important for melanocyte development.

Now some melanocytes don't make it to the left eye. Assuming they weren't going to have blue eyes anyway, the result will be someone with differently colored left and right eyes. The left eye will have fewer melanocytes and so less pigment.

And damage is not the only way to keep melanocytes from traveling and/or surviving. Another way is if someone has something called Waardenburg syndrome.

People with Waardenburg syndrome have mutations in certain genes that might cause some melanocytes to get lost on their way to where they are supposed to go. This means that people with Waardenburg syndrome can end up with an eye that didn't get any melanocytes (and so is blue). Or a patch of hair or skin that ends up white because melanocytes couldn't find their way there.

Sometimes someone can end up with one eye that has two different colors. Because some melanocytes made it to the right place and some didn't.

There are changes in several genes that can cause Waardenburg syndrome. Mutations in any one of them will disrupt the normal development of melanocytes. This isn't surprising given how complex it is to make a cell and have it travel to the right spot. Something that complicated is going to be controlled by lots of genes.

So lost or destroyed melanocytes explains a lot of heterochromia. But probably not all of it.

Another possibility is that each eye has a different set of eye-color genes. There are a couple of ways this could happen.

Eye-color gene versions

Let's focus on the OCA2 gene for this. OCA2 is a key eye-color gene. Certain versions of this gene say to make lots of pigment and so lead to brown eyes. Other versions that say "make no pigment" lead to an eye color other than brown. (These other versions don't necessarily give blue eyes because there are other eye-color genes besides *OCA2*.)

So one way that someone could end up with different-colored eyes is if the *OCA2* genes were different in each eye. For example, imagine that the right eye has a brown version of *OCA2* and so is brown. And that the left eye does not have a brown version and so is green or blue.

As I said, there are a couple of ways this could happen. One is something called chimerism. This is essentially where two fertilized eggs fuse together to create a single person (see the answer starting on page 74 for a more detailed look at chimeras). If each fertilized egg has a different set of eye-color genes, then a chimera can end up with two different-colored eyes.

The other way is something called mosaicism. Here what happens is that a gene gets changed early on in a single cell. All cells that develop out of that mutated cell are now subtly different than the rest of the cells in the body. If that subtle difference is in an eye-color gene, then you may end up with two different-colored eyes.

Both of these conditions result in each eye possibly having a different set of eye-color genes. And so result in a person with two different-colored eyes.

Now neither of these is probably as common as damaged melanocytes. And there are other ways to end up with different-colored eyes too.

Sometimes deposition of foreign material or blood as a result of injury to the eye can cause an iris to look darker. And sometimes the use of certain eye drops can stimulate the melanocytes to make more pigment. Both of these can explain some cases of heterochromia.

So, have we explained David Bowie's eyes yet? Believe it or not, no.

We know his eye color is the result of a fight. But the injury didn't affect melanocytes or deposit blood in the iris. Instead, his eye-color change comes from a permanently enlarged pupil that makes his eye look dark.

The size of the pupil is controlled by muscles in the iris. The muscles are controlled by nerves running from the brain to the eye. Damage to the nerve that normally constricts the pupil produces an abnormally large pupil.

I don't know the explanation for the other famous people I brought up earlier. Nor do I know why actress Kate Bosworth has one blue eye and one hazel eye. Or why lead singer Tim McIlrath of Rise Against has one brown eye and one blue eye. But we've learned there are lots of ways to end up with this rare condition.

Some are genetic like Waardenburg syndrome or chimerism and mosaicism. And some are the result of injury either in the womb or, as in the case of David Bowie, in a bar.

Answered by Dr. Azita Alizadeh, Stanford University

> **My eyes were medium-brown when I was younger, but after I hit puberty they changed color. They are now a really dark brown-hazel. How/why did my eye color change?**
>
> A high school student from Indiana

IT IS AMAZING the number of questions out there about eye color. Eye color is taught as this wonderfully simple genetic trait and then almost everyone quickly comes up with all sorts of exceptions.

Sometimes eye color can appear to change based on surrounding colors. The ambient light and even the color of your clothes or makeup can make the color of your eyes look different. This is most likely to happen with lighter-colored eyes.

But there are also many reports of eyes actually changing color. For example, in one study, 15% of Caucasians had some lightening or darkening in eye color at puberty. In fact, this study showed there was probably some unknown gene or genes involved in the eye-color change.

What is surprising to me is that eye color doesn't change more often. Eye color is determined by lots of different genes, but it all boils down to how much pigment you have in the front part of your iris at any one time. Lots of pigment means brown eyes, a little bit, blue eyes. Other colors come from intermediate amounts of pigment.

The genes involved in eye color determine how much pigment gets made, how quickly it is degraded and where in your iris to put it. In other words, eye color is an ongoing process that is not necessarily set in stone.

So all that has to happen to change eye color is to change the final amount of pigment in your eye. How could that happen?

Remember, genes are just recipes for proteins. When eye-color genes are on, proteins that make and degrade pigment are made. The amount of pigment in your eye is determined by how good these proteins are at their job and how many of these proteins are doing their jobs. For example, you get the same amount of pigment if you make a little bit of a good protein or lots of a mediocre protein.

The most likely explanation for a change in eye color is a change in the amount of pigment-producing proteins made. There are lots of cases where something in the environment changes the amount of protein that is made.

Now, back to your question. An eye-color change at puberty doesn't seem farfetched considering all the genes that get turned on and off when a child turns into an adult. In fact, maybe the 15% of people whose eyes change color at puberty have an eye-color gene that responds to the sex hormones associated with puberty.

As for eyes changing color at various times as an adult, we need to say that there is something in the environment affecting one or more of the eye-color genes. There are lots of examples of things in the environment influencing how much a gene is turned on. Stress, for example, is known to affect genes important for the immune system. I've also read about certain foods affecting eye color.

The bottom line is that eye color is the result of a constant process of pigment creation and destruction. As I was writing this, I began to wonder if most people have small changes in their eye-color genes, but that it is unnoticeable. For example, my blue eyes are most likely due to defective eye-color proteins. So if their expression changed, there would be no change in eye color. The same probably holds true for dark brown eyes where any changes are too subtle to notice. It may be that only those on the cusp with, for example, hazel-color eyes can notice these slight changes.

Answered by Dr. Barry Starr, Stanford University

BLACK OR WHITE OR RED ALL OVER

Questions about hair and skin color

Green is Beautifuul...

Where do different skin colors come from?

A curious adult from California

HUMAN SKIN COLOR can vary from almost translucent to almost black. This range of colors comes from the amount and type of a pigment called melanin found in the skin.

There are two types of melanin—eumelanin and pheomelanin. In general, the more eumelanin in your skin, the darker your skin will be. People who make more pheomelanin than eumelanin tend to have lighter skin with freckles.

Like many other traits, the amount and kind of pigment in your skin are controlled by genes. The versions you have of each of these genes work together to create the final product—your skin color.

To understand how this works, we'll talk about some of the genes that scientists have found that affect human coloring. And how some fish helped us find these genes!

It is also interesting to think about why we have different skin colors. Later on, we'll see how vitamins and where your ancestors lived might have influenced your skin color.

Melanin and your skin

Melanin is made in special cells called melanocytes. These cells are found in the epidermis of your skin.

There are at least four ways people can end up with different skin color. One way is if people make less pigment. Less pigment = lighter skin.

Another way is when people have fewer melanocytes. Fewer melanocytes means less pigment overall and so lighter skin.

A third way is how the melanin is packaged inside the skin cells. The size and shape of the packages affect how the melanin interacts with light, which affects the color.

The fourth way is a bit more complicated and has to do with how much of each of the different kinds of pigment someone makes. There are two types of melanin. Eumelanin is black or brown pigment and pheomelanin is red or yellow pigment.

People who make lots of pheomelanin tend to have fairer skin, because this pigment is lighter than eumelanin.

Melanocytes are usually spread pretty evenly in the skin. Sometimes a few melanocytes will make more melanin than their neighbors. These spots of extra melanin are freckles.

Skin-color genes

Scientists have figured out that several genes are involved in skin color. One of these genes is the melanocortin 1 receptor (MC1R).

When MC1R is working well, it lets melanocytes convert pheomelanin into eumelanin. If it's not working well, then pheomelanin builds up.

Most people with red hair and/or very fair skin have versions of the MC1R gene that don't work well. This means they end up with lots of pheomelanin, which leads to lighter skin.

Two other skin-color genes were first identified in fish. One gene was found in stickleback fish and the other in zebrafish.

Researchers studying the stickleback fish found that the *kit ligand* gene *(kitlg)* was different between dark and light stickleback fish. They also found that humans have different versions of this gene too! And that certain versions lead to lighter skin.

The KIT ligand gene is needed for the survival of melanocytes. So if a person (or a fish) has a version of this gene that doesn't work well, their melanocytes won't survive as well. Fewer melanocytes will mean less pigment. And so lighter skin.

Researchers studying zebrafish with light-colored stripes found another gene involved in human skin color, *SLC24A5*. The fish with light-colored stripes had a version of this gene that didn't work well. When they looked in people, researchers found that some lighter-skinned people also had a poorly working version of this gene.

Unlike the case with *KITLG,* scientists don't know for sure what *SLC24A5* does. Some clues suggest that it may be important for moving calcium into cells. It may be that calcium is important for having more melanin in cells.

Scientists have figured out that lighter-skinned East Asians get their skin color mostly from a non-working version of *KITLG.* Northern European people with lighter skin often have a poorly working version of *SLC24A5.* A number of pale northern Europeans get their skin color from a non-working *MC1R* gene.

Although these three genes help to account for skin color differences between populations, there are probably other genes that scientists have yet to find. And scientists are hard at work to find the genes that make different people within a population have different-colored skin. Think northern and southern Europeans. Or eastern and southern Asians.

Why different skin colors may have evolved

It is pretty obvious that people whose ancestors come from Northern Europe or Japan tend to have lighter skin than people whose ancestors are from sub-Saharan Africa or Australia. The reason for these differences may have to do with the amount of sunlight in each place.

Sunlight can be pretty dangerous stuff because of its ultraviolet (UV) light. UV light can do things like destroy folic acid or cause changes (mutations) in the DNA of some skin cells. Sometimes, these mutations can lead to skin cancer.

Melanin protects skin from damage by sunlight. Since darker skin has more melanin, the darker color is helpful when there is a lot of sunlight.

But sunlight isn't all bad. Sunlight can help our bodies make vitamin D.

Everyone needs vitamin D and you have probably seen that it is in things like milk. When we don't have enough vitamin D, it can cause problems with your bones. Vitamin D deficiency can lead to diseases like rickets or osteoporosis. It has even been linked to some types of cancer.

When our skin gets UV rays from the sun, our bodies use the UV light to make vitamin D. But melanin in our skin acts like a filter, making it harder for people with more melanin (darker skin) to make vitamin D. This means that the more melanin you have, the more sunshine you need to make enough vitamin D.

Imagine you lived somewhere cold and dark. Your skin wouldn't see much sunshine. How would your body get enough sunshine to make all of the vitamin D you needed? Well, if your skin had less melanin (which would make it lighter-colored), then you would need less time in sunshine to make enough vitamin D!

So it's possible that lighter skin evolved so that people in darker places were able to get all of the vitamin D they needed to stay healthy (and have kids).

So the skin color you have might be a result of how much sun your ancestors got!*

Now we know that skin color is just a matter of how much melanin you have. The versions of the skin-color genes tell your body how much melanin to make. All of this means that the difference between dark and light skin is only a few changes in DNA!

Answered by Jamie Conklin, Stanford University

* Of course there are other theories too.

Why do I have freckles?

A middle school student from California

SOME SAY FRECKLES come from the kiss of an angel. But we know where most of them really come from. They come from genes. And the sun.

Freckles are spots usually seen on people with fair or light skin. These spots contain melanin, a chemical pigment.

Melanin gets made when harmful ultraviolet (UV) light from the sun hits our skin. This UV light is very powerful and can damage DNA. Melanin shades and protects our DNA from the sun.

Not every cell in our body makes pigment though. The cells in the skin that make melanin are called melanocytes.

In some people, these melanocytes are spread out pretty evenly in the skin. These folks tan.

Melanin, the pigment, can sometimes collect in the skin cells. This collection of pigment can result in freckles, while the rest of the skin stays light (or turns red) instead of tanning.

Do we know why people with freckles have this pigment collection? We know some of the reason, but not the whole story.

A big clue was the fact that people with red hair tend to also have freckles. And that freckled people often have red-haired children. Why is this?

MC1R

Because most cases of freckles and red hair are caused by the same gene—*MC1R*. Here's how it works.

The *MC1R* gene is really just a set of instructions for making the MC1R protein. The MC1R protein sits in the outer layer of melanocytes where part of its job is to make sure that there is a balance of certain pigments in hair and skin.

There are two main pigments in people. The most common one is eumelanin. It is responsible for most hair colors other than red.

The other pigment, pheomelanin, is more rare. It is responsible for red hair and the orangish-red look of many freckles.

When MC1R is working, it turns pheomelanin into eumelanin. So when MC1R is broken, you get a buildup of pheomelanin. And red hair. So people with broken *MC1R* genes have red hair.

But if MC1R is responsible for both red hair and freckles, why do some people only have freckles? It has to do with how many copies of certain versions of the *MC1R* gene they have.

Remember, everyone has two copies of most of their genes. And these genes come in different versions.

Some people have versions of the *MC1R* gene that don't work well. If one of their copies doesn't work well, they can get freckles. If both don't work well, they get red hair and freckles.

Now the red hair makes sense. If MC1R doesn't work, you'll get a buildup of pheomelanin and end up with red hair. But why does having half your MC1R work sometimes cause freckles?

Scientists don't know. What they do know is that having one or two non-working copies of the *MC1R* gene can sometimes cause freckles. Note the sometimes.

MC1R is not the whole story

Not everyone who is a carrier for red hair has freckles. So there must be more to freckles than the *MC1R* gene itself. We haven't yet found what that other thing is.

Some recent work also shows that there are other ways to get freckles besides just the *MC1R* gene. A group of scientists in China found another freckle gene. This one was on chromosome 4 in a Chinese family with freckles. And two more genes are strongly associated with freckles in Europeans.

So as you can see, we have part of the story on freckles but don't know everything yet. Having broken copies of the *MC1R* gene isn't always enough. And there are other ways to get freckles besides *MC1R*.

Most likely we are missing some other genes needed for freckling. Certain versions of these genes might need to be there along with certain versions of *MC1R* to end up with freckles. Scientists will keep looking.

MC1R is a very interesting gene. It leads to the confusing result that freckles are dominant and red hair is recessive. Even though they are the same gene!

The *MC1R* gene is actually responsible for hair color in many animals. Blond or ginger-colored mice were found to have variations in the same gene. Scientists have also recently found that woolly mammoths that were light-colored also had the same MC1R changes.

Hair, skin, and particularly eye color can all get pretty complex. Most often, these traits aren't as simple as being dominant or recessive, like having red hair or freckles. Our genes interact in many different ways to finally make us who we are.

Answered by Dr. Devasena Gnanashanmugam, Stanford University

Can you tell me more about the genetics of hair color? My husband and I both have dark blond hair, our son has blond hair, but our daughter was born with dark brown hair. I thought that blond hair was recessive. Am I wrong?

A curious adult from Oregon

MAN, I THOUGHT the eye questions were tough! There is very little known about hair color inheritance but there are some interesting theories. There is currently a lot of research on the genetics of color so we are sure to learn more soon.

What is pretty well known already is where hair color comes from. Hair color happens because of a kind of pigment called melanin. There are two kinds of melanin, eumelanin and pheomelanin.

For the sake of ease, we'll first focus only on eumelanin. If your hair has a lot of eumelanin, it will be black; a little eumelanin and your hair will be blond.

The amount of eumelanin in your hair is determined by lots of genes. Let's imagine (although the real case is more complicated) that there are two possibilities for each of these genes, either on or off. When a gene is on, it helps make eumelanin and when it is off, it doesn't make anything.

One other thing you need to know is that genes important for making eumelanin work in an additive way instead of in a dominant and recessive way. So the more genes that are turned on, the more pigment gets made and the darker the hair. Having more genes turned on is like adding more dye to a bucket of paint. Each "on" gene adds a bit more dye.

Given these assumptions, the answer to your question is that your son inherited few of the "on" eumelanin genes while your daughter inherited a lot.

To put the answer in more concrete terms, we'll imagine that there are four eumelanin genes that determine hair color. Remember you get one copy of each gene from your mother and one from your father giving you a total of eight copies of these hair-color genes.

If one of these hair-color genes is on, we'll represent it with H and if it is off, we'll represent it with h. Using this system, someone with very black hair would be $HHHHHHHH$ and a blond person would be $hhhhhhhh$.

You said both you and your husband have dark blond hair. If we imagine that both of you are $HHHhhhhh$, then it is easy to imagine how your son's and daughter's hair color came about.

Remember, your kids get one copy of each of the four genes from each of you and there is no dominance per se—they add up to give a hair color. If each of you contributed only h's, then you would get a blond-haired kid like your son with a genotype of $hhhhhhhh$. Your daughter got more H's than h's (perhaps $HHHHHhhh$) and so has dark hair.

Blond– vs. brown–hair genes

So what are the genes that make hair blond or brown in Europeans? Scientists have been looking at the DNA of thousands of people to find out.

Researchers studying Icelanders and the Dutch found clues to six big sections of DNA that might be important in determining brown- vs. blond-hair color in Europeans. They looked more closely at some known coloration genes that are in these regions.

Some of the genes they found are already known to be important for skin color, freckles, or eye color. They include genes such as *MC1R, OCA2,* and

KITLG. It makes sense that freckles and hair, eye, and skin color have some genes in common since they all involve melanin.

Versions of genes that cause less melanin to be made in skin or in the eye might also make less melanin in hair. And then again, they might not. Scientists are still working through these genes to figure this all out.

The researchers also found some genes that were not known to be related to color. Scientists don't know what these genes do. Their best guess is that at least some of them are important for moving calcium into cells. Calcium might be needed to produce more melanin.

These genes are just some of the genes that have to do with making hair blond or brown. There are probably more genes that scientists haven't found yet. Natural blonds are found in many populations around the world. Different genes could make hair blond in different ethnic groups.

Whatever the complete set of blond vs. brunette genes turns out to be, the genes will work together to determine the blond or brown shade of hair.

Answered by Drs. Barry Starr and Dale Bodian, Stanford University

> ## Why does hair lighten in the sun but skin darken? Does it have to do with melanin?
>
> A curious adult from Brazil

THAT IS A GREAT QUESTION and worth investigating! You are right in thinking that melanin has something to do with the answer. Melanin is a pigment found in your skin and hair cells that gives each its color.

It does seem weird that the sun bleaches our hair and darkens our skin. This mostly has to do with hair being dead and skin being alive.

The sun bleaches and destroys the melanin in your hair giving you lighter hair. Since hair is dead, your hair will stay that color until new hair comes in.

When sun shines on your skin, it destroys the melanin as well. But since your skin is alive, it can respond to the sun's damage. Your skin cells make more melanin and your skin becomes darker. That's the simple answer, now let's get into the details.

Special cells in the skin, called melanocytes, make two different types of melanin, eumelanin and pheomelanin. Eumelanin has a brown or black color while pheomelanin is yellowish-red. How much of each kind of melanin we make determines the exact shade of our skin and hair color.

Melanin provides UV-damage protection

Why do we even need this coloring? Melanin helps protect our skin and hair by filtering out potentially harmful ultraviolet (UV) radiation from the sun.

The UV light from the sun can damage DNA and cause cancer. This is one reason why our body darkens the skin when exposed to sun—to protect us.

Skin-darkening is a two-step process. First our cells experience the "immediate pigment darkening" response when exposed to the sun. This happens when our cells are exposed to UVB or ultraviolet type B radiation.

This is a quick response that happens over a period of minutes to days. You can't easily see this happen with lighter-skinned people but it is more obvious with people with darker skin.

The second and slower response of our skin cells to sunlight is called "delayed tanning". As the name suggests, it's much slower. This second response is also called melanogenesis and happens when our skin is exposed to UVA radiation.

Melanogenesis just means the cell is ramping up to make more melanin pigment. Part of this is that more melanocytes get made. Also, melanin genes start making more melanin so that the melanocytes are darker. So you end up with darker melanocytes and more of them.

Strangely the sun is not directly causing more melanin to get made. It's the damage that the sun is doing to the cell that starts this process.

When the sun damages a skin cell, the cell releases chemicals alerting the body that it has taken a hit. These chemicals cause more melanocytes and more melanin to get made.

Of course, this process needs to stop at some point. If it kept on, we'd call melanogenesis skin cancer. So the body stops the process with some different chemicals and we eventually turn lighter for the winter.

None of this can happen in hair because hair is dead. So our hair has to take the full brunt of the sun without any defenses besides the melanin it was originally made with.

And sunlight is pretty energetic. For example, look how it ruins plastics—like vinyl car roofs. In time, sunlight will destroy some plastics causing them to fall apart and crack. In chemical terminology, the ultraviolet radiation in sunlight ruins things by oxidizing them.

Similarly, the UV in sunlight oxidizes melanin into a colorless compound. This is why your hair gets lighter. Blond hair is really just colorless hair or hair with very little melanin in it.

But melanin isn't the only protein taking a beating in your hair. The sun is also clobbering other proteins making hair less manageable as well.

More sun damage to hair

There is a chemical group in hair called a thiol. When thiols are unoxidized, the hairs slide easily across each other. When the sun oxidizes those thiols, it is a whole different story.

When a thiol group is oxidized to a sulfonic acid, the hairs tend to stick together more. So you get tangling that can't be repaired.

Once the thiols on hair have been oxidized to sulfonic acids, there's really no going back. All you can do is to treat the hair with conditioners and wait for new hair to grow in.

So, tanned skin and bleached hair may be a sign that someone is spending too much time in the sun. Try to keep your skin light and your hair dark by wearing sunscreen and a hat!

Answered by Dr. Aaron Shafer, Stanford University

THE LONG AND THE SHORT OF IT

Questions about earlobes and other physical traits

Are baldness and other generally male traits due to dominant genes on the Y chromosome, or recessive genes on the X chromosome?

A middle school student from Alabama

EXCELLENT QUESTION! In fact, both possibilities happen. Becoming a male is due to a dominant gene on the Y chromosome. But many of the special problems males have are because of recessive genes on the X.

Let's quickly go over what dominant and recessive are to make sure we're on the same page. Remember, we have two copies of most of our genes— one from mom and one from dad. And these genes come in different versions, which are called alleles.

These versions can be either dominant or recessive. If you only need one copy of a gene version to get a trait, then it is dominant. If both copies need to be the same, then the trait is recessive.

For example, a big part of eye color is determined by the *OCA2* gene. This gene comes in two forms—brown or blue.

You only need one of your copies to be the brown kind to have brown eyes. But to have blue eyes, both copies need to be blue.

Okay, now let's see how dominant and recessive genes relate to being male.

Sex chromosome genes in males

As you may already know, males have an X and a Y sex chromosome and females have two X chromosomes. This means a couple of things.

First, there is something about having the Y chromosome that makes a person male. And second, males have only one copy of all of the genes on the X and the Y chromosomes.

There is a gene on the Y chromosome called *SRY*. This gene tells a fetus to develop male body parts. It is a dominant gene.

There are cases where someone with two X chromosomes also has an *SRY* gene. These people look male. And there are cases where someone has an X and a Y chromosome but the *SRY* gene doesn't work properly. These people look female.

So becoming male is due to the dominant gene *SRY*. Anyone who gets one working copy of this gene looks male.

The other important consequence of being male is that males have only one copy of the X chromosome. This means that they only have one copy of all of the genes on that chromosome.

So what this means is that it only takes one recessive gene on the X chromosome to see a recessive trait in males. If the eye-color gene were on the X chromosome, more men would have blue eyes.

This explains why males tend to be colorblind more often than females. And why they also get hemophilia, a blood clotting disorder, more often and they suffer worse from fragile X syndrome. And why they also go bald more often.

All of these traits come about because of a recessive gene on the X chromosome. Men only have one copy of these genes because they only have one X chromosome. So they suffer more often from these recessive traits than females do.

At least this is certainly true of most of these conditions. Balding is an exception though.

Balding is not a simple trait

There does seem to be an important gene on the X chromosome that causes men to bald early in life. But it is not the only gene involved.

The kind of balding we're talking about is male-pattern hair loss (MPHL). This common type of baldness generally occurs in males. It can also occur in females but this is rare.

So how does this work exactly? Why do men get MPHL more often than women? It isn't as simple as colorblindness or hemophilia.

The scientific name for this kind of balding is *androgenetic alopecia*. "Andro" refers to androgens. These are hormones that our bodies produce, like testosterone and dihydrotestosterone (DHT). "Genetic" means that there is a gene or many genes that need to be inherited to get MPHL.

So this type of balding is the result of genes and androgens. Our bodies use androgens to turn on certain genes. To do this they need a gene on the X chromosome called the androgen receptor. Well, to be more precise, they need the protein the gene codes for—the androgen receptor (AR).

AR grabs testosterone or DHT, then heads into the nucleus where most of our DNA is, and turns on certain genes. We know that males who have too many androgen receptors tend to be bald.

In 2005 scientists performed a major study that has brought us closer to finding out what is going on with MPHL. They took several hundred balding and non-balding men and compared their androgen receptor *(AR)* genes.

The researchers found that the balding men were more likely to have a certain version (or allele) of the *AR* gene than the non-balding men. This means that one factor for determining MPHL is the X chromosome a man got from his mother. But this is unlikely to be the only factor and there are probably many other genes that cause MPHL.

It does help explain why baldness is linked to mom's side of the family. And why men are more often bald than women.

But, as I said before, the *AR* gene is probably not the only gene involved. There are likely many more factors and some will probably be on one of the non–sex-linked chromosomes, or autosomes. Each parent randomly contributes these genes so you are as likely to get them from your dad as from your mom. In fact, there is evidence that chromosome 20 may be involved.

So scientists are still trying to figure out the genetics behind MPHL since it's not as simple as one gene. There could even be environmental factors that play a role. But for now, one of the best predictors we have for baldness is which version of the *AR* gene a male gets from his mother.

So there you have it. Being male is the result of a dominant gene. But many of the problems that plague males more often than females are due to recessive genes on the X chromosome.

Answered by Monica Rodriguez, Stanford University

> **How can two parents with dwarfism have a child that is of average height?**
> A high school teacher from Pennsylvania

THIS IS A GREAT QUESTION. It can be hard to think about how this could happen. If the parents have dwarfism, the kids should too, right?

Well, no. This is only true of recessive traits. Dwarfism is most commonly a dominant trait.

Very useful answer, huh? Let's dig a little deeper to find out what the answer really means.

We have two copies of most of our genes—one from mom and one from dad. And our genes can come in different versions called alleles.

For example, there is a gene called *FGFR3*. Most people have versions of this gene that result in average height. But some versions can lead to dwarfism.

People with dwarfism have one copy of *FGFR3* that causes dwarfism and one that does not*. In genetics speak, this means that the dwarfism version is dominant to the other, more common version.

Now imagine two parents with dwarfism. Which copy of a gene we get from our parents is random. So each parent has a 50-50 shot of passing down a copy that does not lead to dwarfism.

---

* If both copies lead to dwarfism, then the baby will be either stillborn or die shortly after birth.

What this means is that 25% of the time, the child will get a copy of the average-height gene from both parents. The result will be a child of average height.

So that explains it. Two parents with dwarfism can have a child of average height because the parents are carriers of that trait.

But wait a minute. If dwarfism is dominant, where does it come from in the first place? Do people with dwarfism always have little people as parents? No.

The majority of cases of dwarfism result from DNA changes in the *FGFR3* gene that appear out of nowhere. Literally. These changes (or mutations) happen in the sperm or the egg or very early on in development.

DNA can change

This brings up the very important point that genes are not written in stone. They are written in DNA. And DNA can change.

DNA is made up of four letters that spell out, in three letter words, the instructions found in a gene for making a protein. A mutation happens when something in that series of letters changes.

Either one or more letters go missing, an extra letter or more is added, or the letters are changed to a different letter. These changes alter the instructions giving a different version of the protein.

Where do these changes come from? DNA can change because of something in the food we eat or the air we breathe, from sunlight, or a host of other environmental factors.

Or our cells can change the DNA accidentally. This happens occasionally when our cells copy our DNA. The copying process is incredibly reliable, but still our cells occasionally put in the wrong letter. All the cells that come from that cell will now have that mistake.

Incidentally, mistake may be the wrong word to use. These "mistakes" are the stuff of all of the wonderful variety we see around us. They are responsible for big things like evolution and for little things like red hair, blue eyes, or dark skin. Thank goodness genes are made of DNA and not stone!

So there you have it. Two parents with dwarfism can have a child of average height because dwarfism is a dominant trait.

And the parents probably did not inherit their dwarfism from their parents. At some point early in development, their *FGFR3* gene picked up a DNA change that led to dwarfism.

Answered by Dr. Barry Starr, Stanford University

> **How it is possible that I ended up with one earlobe attached, and the other unattached? I have read all about the dominant gene process etc., but it never seems to address how you end up with one of each.**
>
> A curious adult from Washington

I HAVE A FRIEND who has an attached and an unattached earlobe too. Very cool. But how in the world can something like this happen?

As you say, unattached earlobes are thought to be dominant over attached ones. So if one ear is unattached, they should both be unattached. And vice versa.

But what may have happened for you is that one ear followed the dominant instructions and the other followed the recessive ones. In other words, each ear had a different set of instructions.

One way this could happen is if the DNA is different for each ear. There are a couple of ways this can happen.

Chimeras

One is if you are a chimera. A chimera is someone who has two groups of cells in their bodies, each with their own DNA.

The differences between the two DNAs are huge. As different as two siblings.

And this makes sense because a chimera is the result of a fusion of two siblings. What happens is that two fraternal twins fuse very early on in development*.

Now, this new combined embryo grows into a single person. Some of their cells are the DNA of one sibling and some are of another sibling.

Imagine one twin was going to have attached earlobes. And the other was going to have unattached earlobes. Fuse the two and you might end up with someone with an attached and an unattached earlobe.

This isn't the most likely explanation since chimeras are thought to be relatively rare. As of 2003, there were only 30 documented cases. But this number may be an underestimate because no comprehensive study has been done to figure out how many there actually are.

Mosaics

Another much more common way to end up with two different DNAs is something called mosaicism. This isn't as bad as it sounds. We are all most likely mosaics.

Mosaics happen when a single cell gets a DNA change very early in a fetus' development. That change is then passed down to some cells and not others. The end result is someone with two slightly different DNAs.

Imagine that as an embryo, you are growing and dividing. Suddenly, a change happens in one of your cell's DNA. That change could have come from the environment or simply been a mistake that the cell made copying its DNA.

---

* This happens at a time when the embryo is made up entirely of embryonic stem cells which is why they don't end up with four arms, two heads, etc.

Wherever it came from, let's say the DNA change happens in the earlobe-attachment gene. And the hit changes the instructions from unattached to attached.

You now continue to grow and divide. Each of the cells divides many times to get to the 100 trillion or so cells that make up you.

Now some of your cells have instructions for attaching an earlobe and some cells have instructions for leaving it unattached. If the change happened after the embryo has started making ears but before the earlobe has attached or remained loose, then you might end up with an attached and an unattached earlobe.

But how likely is such a precise change to a certain gene? More likely than you might think if you are converting a dominant version into a recessive one.

This is because recessive genes are often versions of a gene that don't work. So all our hit has to do is garble the instructions, not change them in a precise way.

There are at least a couple of different ways to end up a mosaic. One way changes the actual instructions while the other changes how they are used. The first is more or less permanent while the second is reversible.

What does this mean? The instructions in our DNA are written in a language made up of four bases, A, G, C, and T. These four bases form three-letter words that make up a large part of the instructions for making you.

The type of mosaic we talked about earlier changes one of these letters. For example, imagine that the earlobe instructions are this sentence: "The old man had one new hat."

The mosaic has this intact sentence in some of his or her cells. In the other cells, a letter goes missing and you end up with: "Tho ldm anh ado nen ewh at." A bunch of gibberish that leads to an attached earlobe.

The second kind of mosaic would be one that leaves the letters intact but gets rid of the punctuation. So our sentence goes from "The old man had one new hat." to "the old man had one new hat." In other words, it loses its capital letter.

What happens in this case is the body doesn't know a new sentence has begun so it doesn't know where to start. This is the same as the gene not being there. In our case, it would lead to attached earlobes.

Of course our genes don't really have punctuation. But what they do have is something called methylation.

Methylation

Methylation hides the start of genes by adding a methyl group to the DNA. The bases are the same, but the DNA is used differently.

For methylation to work in your case, we would need it to happen in one set of cells and not in others. The capitalization would be lost only in some cells.

So a way you can end up with a dominant and a recessive trait is if you have two different sets of instructions.

Of course, we can be pretty sure there are other possible explanations as well. Perhaps some sort of trauma in the womb that causes only one earlobe to unattach. This would be similar to how some people end up with two different-colored eyes (see the answer starting on page 33 for details about heterochromia).

Answered by Dr. Barry Starr, Stanford University

> **I understand how a set of parents *Dd* and *Dd* could have a child that is dimpleless, but is it genetically possible for a set of parents who do not have a cheek dimple or chin dimple to have a child who has both of these traits?**
>
> Anonymous

YOU ARE RIGHT in that it is easy to see how two *Dd* parents could have a *dd* child; each parent needs to contribute a *d* gene. *D* is the version of the dimple gene that gives dimples and *d* is the version for no dimples. Because *D* is dominant, the *dd* child won't have dimples even though both parents do.

The *d* form of the gene is probably recessive because of a mutation. Often in these dominant-recessive gene pairs, the recessive copy of the gene is mutated so it doesn't work anymore which is why the dominant version wins out.

Sometimes, though, you can get a working copy of a gene that doesn't work in you for some reason but does work in your children or grandchildren. This concept is called "variable penetrance" and cleft chin is a classic example.

How does penetrance work? Two well-characterized ways that penetrance can work are: environment and modifier genes.

Environment

Sometimes the environment can influence whether a gene gets expressed or not. Let's look at cleft chin as an example of how this might work.

For some developmental features, a gene may only be on for a short while to cause that feature. Maybe the cleft-chin gene only needs to be on for a short period while the fetus is growing. If something in the environment affects the gene during this short period, it will look like the gene isn't there. If there is no change in the gene's sequence, then the children of this person can have a cleft chin.

Modifier genes

Another way variable penetrance can come about is through modifier genes. Modifier genes are simply genes that affect the expression of other genes.

What this means is that now the appearance of a feature like a cleft chin is dependent on two genes, the modifier gene and the cleft gene itself. If both copies of the modifier gene lead to no cleft chin, then the cleft-chin gene won't matter—it will be silenced. If the modifier gene lets you have a cleft chin, then you still need to have a working copy of the cleft-chin gene.

Clear as mud, huh? Let's try an analogy to hopefully make it a little simpler. Let's compare the situation to the electricity in your house. We can think of the breaker switch, which controls all of the electricity in your house, as the modifier gene.

We'll compare the cleft-chin gene to a lamp in your house. Now, if the breaker switch is off, it doesn't matter whether the lamp is on or not, there's no electricity so it won't work. This is the same situation as with the modifier and cleft-chin gene. If the modifier gene is "off", then it doesn't matter whether the cleft gene works or not—the "lamp" won't work because there is no juice.

So what we need to propose in your case is that both parents have lamps that work but their breaker switches are off. In the kids, the breaker switch and the lamps are on so you get light (a cleft chin).

What might this look like with "real" genetics? Let's suppose that there is a modifier or silencer gene (*M*) that, when dominant, doesn't allow the cleft-chin gene to work (the breaker switch in the off position is dominant). Now, for the example you gave, we need to propose that both parents are *Mm* (they have one working copy of this modifier gene each) and that at least one of them has a working copy of the cleft-chin gene (*C*).

Let's say both parents are *MmCc*; they have a working cleft-chin gene but the *M* masks its presence. There would be a 3 in 16 chance of having either an *mmCC* or an *mmCc* child that would have a cleft chin. A Punnett square (with the three possible children with cleft chin shown in light grey) of this is shown below:

DAD

		MC	*Mc*	*mC*	*mc*
MOM	*MC*	*MMCC*	*MMCc*	*MmCC*	*MmCc*
	Mc	*MMCc*	*MMcc*	*MmCc*	*Mmcc*
	mC	*MmCC*	*MmCc*	*mmCC*	*mmCc*
	mc	*MmCc*	*Mmcc*	*mmCc*	*mmcc*

Since cleft gene is known to have variable penetrance, it is more likely to happen by the modifier gene mechanism.

For cheek dimples, the dimple gene might be masked by a similar penetrance mechanism or may appear in the children by some other mechanism.

As you can see, genetics can get pretty complicated. Of course there could be easier explanations, like the parents had dimples when they were young but grew out of them (which has been known to happen). Hope this helped.

Answered by Dr. Barry Starr, Stanford University

REAL-LIFE CSI

Questions about personal identification with DNA

> ### On CSI they used a medical term for a person that had two different DNA's. Have you ever heard of this?
>
> A curious adult from Alabama

WHAT YOU ARE THINKING OF is "chimera." In the TV show CSI a woman claimed a man raped her, but DNA taken from his blood did not match the DNA of the suspect. The test also revealed that the most likely suspect was a relative of the man. When further DNA tests cleared his relatives, the man's DNA was tested again. This time it was from a hair sample and this time it was a perfect match to the suspect's DNA. Can one person have two types of DNA in different parts of his body?

Yes, this can happen. People with two types of DNA are called chimeras after a mythical creature with a lion's head, a goat's body, and a serpent's tail. We use the word today to describe the same basic idea.

A person who is a chimera is made up of cells from different people. So some of their cells have the DNA of one person and the other cells have the DNA of another person. How do you do that?

Imagine that a woman releases two eggs instead of one. If both of these eggs get fertilized by two different sperm cells, you get fraternal twins. Fraternal twins are no more related than any brother or sister.

Now, imagine that instead of developing separately, these fertilized eggs actually fuse. Then only one baby would develop. This baby would have cells from not one, but two different zygotes.

Remember, since a zygote is a cell that results from the fusion of a sperm cell with an egg, every zygote carries its own unique set of DNA. So the baby would have two different sets of DNA—this baby would be a chimera.

Most chimeras grow up and look like everyone else. You probably wouldn't be able to tell that the person was a chimera. Sometimes you can see patchy areas on the skin or two different-colored eyes. Even these are pretty subtle changes though.

This seems weird at first. Shouldn't a fusion of two people look pretty different? Two heads or four arms or something?

The reason this sort of thing doesn't happen is because the fusion of the two eggs happens very early on in development. At this point, the cells haven't started to build body parts.

After the fusion, a given cell can't tell that its neighboring cell actually has a different set of chromosomes. They work together as if they were all created from the same zygote.

Chimeras are not the only people who carry different sets of DNA in their bodies. Mosaics also have variation in their DNA from one cell to the next.

DNA mutations lead to mosaicism

A mosaic, unlike a chimera, starts out with the same set of DNA in every single cell. You could look at any cell in the body and the DNA inside of it would be exactly the same as the DNA inside a different cell.

At some point during a mosaic's life, though, his or her DNA changes in some but not all of the cells. Now, the DNA in one cell is slightly different from the DNA in the neighboring cell.

This scenario is actually very common. In fact, we are all actually mosaics.

Our bodies are made up of trillions of cells. Most of these cells contain the exact same copy of our DNA. Over the course of our development and lives, however, the DNA in some of these cells can change.

There are many things in the environment that can actually change your DNA. The sunlight hitting your skin and chemicals in the food you eat are two examples.

Most of the time these mutations are harmless. Only occasionally do these mutations cause problems. For instance, if a mutation affects how fast a cell can grow, the cell can become cancerous.

The changes that occur due to environmental factors are usually very small and very rare. Most of the time, only one or a few cells in your body will carry any single mutation.

It is possible, though, for a mosaic to have as many as half of the cells in his body that are different from the other half, just like in a chimera. How?

If the DNA changed very early on in your development, then more cells in your body are likely to carry the mutation. Remember, we start out from a single cell. That cell divides and then each of the two new cells divides. This happens over and over until you end up with 100 trillion cells or so.

Imagine that the first cell has divided to make two cells. During this division, a mistake was made and one cell ends up with a DNA change. Now these two cells go on to make the rest of the person. In this case, half of the person's cells will have different DNA than the first cell had.

Even with these mutations, our own cells are much more similar to each other than they are to the cells from another person. This means that if you were to compare one of your cells to a cell from your brother or sister, you would find a lot of differences. You would find far fewer differences if you instead compared two of your own cells to each other.

So chimeras have cells with vastly different DNA, like a brother's or sister's. A mosaic has cells with only small changes. Does this sound confusing? Let's use the CSI crime scene scenario to demonstrate this difference.

The police had DNA from the crime scene and compared it to DNA from blood from the man accused of the crime.

The test result suggested that the true criminal was a relative of the accused man. Because the accused man was a chimera, though, the DNA may be as different as a brother's. And DNA tests are certainly good enough to distinguish between brothers.

Therefore if the suspect is a chimera, he could get away with murder. (As long as the police don't look at the DNA from another part of his body, like hair.)

But if the suspect were a mosaic, most likely the DNA of two cells will be similar enough that the DNA test would show he did it.

Answered by Natalie Dye and Eszter Vladar, Stanford University

> **Would a person who receives a bone marrow transplant have two sources of DNA? The original in the body's cells and another from the donor cells of the bone marrow?**
>
> A curious adult from Texas

YES. A bone marrow transplant turns the patient into a chimera. What I mean is that the DNA in their blood is different from the DNA in the rest of their cells.

In theory, this could complicate a criminal investigation. In fact, there is at least one case where it did. Before going into that, though, let's go a bit more into how DNA testing and bone marrow transplants work.

DNA testing

DNA has been used in the courtroom as evidence for the last 25 years or so. It has helped convict the guilty and set free the innocent. And it isn't just for the courtroom.

DNA testing (also called DNA typing) can be used to figure out who the father of a child is. It can also be used to figure out fertility issues. Or to figure out if someone is a match for a bone marrow or organ transplant. DNA testing is everywhere!

DNA typing is based on the fact that every cell in a human body contains identical DNA and that everyone's DNA is unique. DNA testing can be done by collecting DNA from very small amounts of human hair, bone, skin tissue, saliva, semen, and blood.

Every person differs from each other in around 0.5% of their DNA. Scientists have identified 13 places on our DNA that tend to be different between individuals. They use those areas to produce profiles that can distinguish one individual from another.

The way these tests are usually done, though, a person's DNA has to be the same in every cell. But this isn't always the case.

Sometimes people are born with two different sets of DNA. These folks are called chimeras (see the answer starting on page 74 for a more detailed look at chimeras).

And sometimes a person develops different DNA later in life. A common example is small DNA changes that happen in some of our cells over time. These folks are called mosaics.

It is also possible to artificially end up with different DNA in some of our cells. Some ways are temporary (like blood transfusions). But others like bone marrow transplants are permanent.

Bone marrow transplants

A bone marrow transplant is used to treat a number of illnesses. It can treat people with various blood and bone marrow diseases. It is also used in the treatment of some forms of cancer.

The way it works is a doctor first destroys a patient's blood cells or bone marrow. This is often done with chemotherapy or radiation. The doctor then puts in new bone marrow from a matched donor.

OK, interesting but why would that affect a DNA test? Because the new bone marrow cells have the donor's DNA. And bone marrow contains blood stem cells. These blood stem cells are responsible for making our blood.

Our blood cells need to be replaced constantly (this is why a blood transfusion only temporarily changes the DNA profile of our blood). What this means in a bone marrow transplant patient is that his or her blood comes from the donor's stem cells. And so has the donor's DNA.

It used to be that patients had all of their bone marrow destroyed. Which meant that the donor's bone marrow completely replaced theirs.

Recently some bone marrow transplant patients get lower doses of chemotherapy and radiation that don't kill all of their bone marrow cells. These patients will have some of their own bone marrow and some of the donor bone marrow. This means their blood DNA profile will be a mixture of both the donor and recipient.

Theoretically this is all fascinating. But has it ever affected a real case? Yes.

A real-life example

Abirami Chidambaram of the Alaska State Scientific Crime Detection Laboratory in Anchorage gave details about one such case. The case involved a serious sexual assault.

Semen was collected at the crime scene and the semen DNA matched a blood sample from a known criminal in the database. But this person whose blood matched the semen was in jail when the physical attack happened. At the same time the crime scene sample also matched the DNA profile of another person.

At first all the detectives were confused by this case. With good detective work they found out that both people had the same last names and were brothers.

They discovered that the person who was in jail received bone marrow from his brother several years earlier. So his blood DNA profile was the same as his brother's blood DNA profile. But his cheek swab DNA profile was different from his brother's.

This case shows that it is very important to test both blood and another tissue in a suspect's body to make sure they show the same DNA profile. So police may have to check both blood and cheek samples to be sure of recognizing a transplant recipient. Or even a natural-born chimera.

This case also points out the small risk that potential marrow donors take by having their DNA profile turning up in a crime database if the recipient later commits a crime. But this risk is probably better than the alternative.

Answered by Dr. Azita Alizadeh, Stanford University

> **If DNA evidence at a murder is from an identical twin, can scientists tell which twin killed the person?**
>
> A curious class and a curious adult from California

IT SEEMS LIKE a police officer doesn't have a shot in this case. Identical twins after all are identical. They start out with the same DNA.

But DNA isn't a constant thing. DNA builds up small differences over time because of the environment. We all build up mutations in our DNA over time. Most of these DNA changes are harmless, although some can lead to diseases like cancer.

Where do these changes come from? Some come from the stuff our body does every day. For example, we all start out with a single cell and end up with somewhere around 50 or 100 trillion cells.

The DNA in all of these cells needed to be copied (not 100 trillion times but a lot). The machinery in our cells that copies our DNA is incredibly good at what it does, but not perfect. Occasionally, it makes a mistake that is not fixed.

Our DNA also changes in response to things like sunlight or the food we eat. Both can damage the DNA causing mistakes to happen.

What this all means is that identical twins have differences in their DNA. They are not as identical as we all think. But that doesn't mean there are easy genetic tests that can tell them apart. There aren't.

The main problem is that most of the changes aren't in all of your cells—not all of your cells have the same DNA sequence! If a DNA mistake happens late in our development, then only a few cells will have that mutation. If a mistake happens early, then more cells will have the DNA change but still not all of them. These changes can even happen in adults!

To work, the genetic test is going to have to sequence a lot of DNA to find the few base pairs difference between the twins. Not only that, the police will need to know what tissue the DNA comes from so they can compare it to the right tissue in the twins.

But mutations aren't the only way DNA changes. Another way to tell identical twins' DNA apart has to do with chemical changes on the DNA. These changes tell a cell how often a gene should be read. These markings change as we age in ways dependent on our life experience.

This means that identical twins have different DNA markings on their genes. Scientists have recently figured out how to quickly and easily figure out how much of one of these marks, methyl groups, are on a bunch of genes all at once. Someday, perhaps in the not too distant future, scientists will be able to use this method to catch a murderer. But it won't be easy!

Genes can be on, off, or in between

Genes are an important part of our DNA. Each gene has the instructions for making a protein and each protein has a specific job it does in the cell.

Not all genes work in all cells though. For example, our skin doesn't need a protein that helps us see. So the gene responsible for this protein is off in our skin and on in our eyes.

Genes can also be turned up or down. In other words, a gene that is on is not always on at the same level. Sort of like a dimmer switch on a light.

What this means is that there are differences between how much genes are turned on between different kinds of cells. And between different people too. Including identical twins.

Identical twins use their genes differently

Scientists have found that when identical twins are born, they use their genes pretty similarly. For example, the genes on in blood cells are pretty much at the same level between the two twins. The same is true for all of their different genes and cells.

As the twins age, though, this pattern changes. Now some genes are on higher or lower in one twin compared to the other. These changes reflect the different life experiences of each twin.

One reason for some of these changes in how genes are used is the methylation I talked about earlier. Methylation is really just how many methyl groups are stuck on or near a gene. And a methyl is just a small chemical group.

Lots of methyl groups tend to shut off a gene while having a few means the gene is on. So two genes with the exact same set of DNA letters might have different levels of methylation in each identical twin. This difference can be used to tell one twin from the other.

Recently scientists have come up with ways to quickly figure out how methylated lots of different genes are all at once. This might be a way to tell identical twins apart from DNA evidence. And so catch our murderer.

So maybe scientists could look at the DNA left at a crime scene and match its methylation pattern to one twin or the other. This would require that the crime be recent since methylation patterns do change.

The bottom line is that there are definitely differences between the DNA of identical twins. Finding these differences will be hard but it is getting easier all the time.

This isn't just a thought experiment. This is a real problem the police have come up against. In fact, in a case in Michigan, the police haven't been able to tell which twin brother committed a crime because there was no eyewitness or fingerprint. Soon, we may be able to put the guilty identical twin in prison with just DNA evidence.

Answered by Dr. Barry Starr, Stanford University

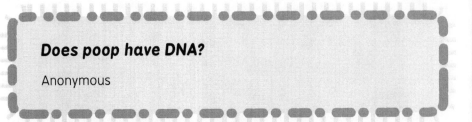

Does poop have DNA?

Anonymous

WHEN SCIENTISTS STUDY a population in the wild, they don't want to disturb the animals too much. So how can you study animals without capturing them? One way is by looking at their poop.

Think about it. You spend ten years in college and graduate school and what do you do? You go to the woods and collect poo.

And it is really important work. Poop has DNA. And every individual has a unique set of DNA. This goes for people, chickens, fish, and most anything that doesn't reproduce by cloning.

What this means is you can tell one tiger's poop from another's by looking at the DNA in it. So you can figure out how many different tigers (or any other land animal) are living in one place. You can also tell what bacteria or viruses an animal might have by looking for viral or bacterial DNA.

Doing things like this, scientists figured out that there might be more pandas in the wild than they once thought. And from what group of wild chimpanzees HIV originated. All from DNA in poop.

And this doesn't even mention other things scientists have learned. For example, grasses apparently evolved much sooner than scientists originally thought. How'd they figure this out? From grass in dinosaur poop.

How many pandas are there anyway?

Deciding whether an animal is endangered is important work. One of the ways we figure this out is by seeing how many individuals of a certain species are left.

But how do you do that? Usually by counting the animals you can see. You can trap and tag them. Or if they're fish, force them through a counting station. Or just watch with your eyes or a camera and see how many animals you see.

Not a very precise science! What do you do with shy ones? Or smaller ones? Or nocturnal ones? How many are killed in a trap? Which ones do you trap? Just the stupid ones? Or light-colored ones? Or ...

Believe it or not, a lot of these problems can be solved by looking at an animal's poop. You may never see a panda but it will leave poop where it has been. And if the poop is fresh enough, the DNA can be looked at.

As anyone who watches CSI knows, every person's DNA is unique (except identical twins). If you do enough testing, you can figure out the difference between two people's DNA. Same thing goes for pandas.

What some Chinese scientists did was to collect a lot of fresh panda poop. They got the DNA and made genetic profiles of each of the pandas in a reserve.

Using conventional counting, scientists had estimated that around 32 pandas were living in the reserve. By studying DNA, the researchers concluded that somewhere between 66 and 72 pandas were actually there. More than twice as many as was previously thought.

This is great news if it can be extrapolated to all of the estimated 1600 pandas thought to be alive in the wild. Notice the "if". We don't know yet if the same thing will be true in other places.

And while this is great news and good science, it is not cheap. It would be very difficult to do this for every endangered species. Imagine the outrage at scientists paying millions of dollars to collect and study an endangered lizard's poop. No, this will probably only be acceptable for cute, cuddly, high-profile beasts like the panda.

But still this is a great use for poop DNA. And may prove useful for other endangered animals as well.

Where did the AIDS virus come from?

AIDS seemed to appear out of nowhere at the beginning of the 1980's. After some work, scientists concluded that it originated somewhere in Africa.

But why hadn't we seen it before? How does a virus just pop up and start wreaking havoc? One way a virus can just appear is when it jumps from one species to another. This has been a big worry with diseases like the avian flu.

Avian flu is a virus that was killing lots of birds but only a few people when it was first identified. It wasn't killing many people because it couldn't go from person to person—people only got it from close contact with birds.

The fear was that the virus might mutate and gain the ability to move from person to person. And the fear is a real one.

Viruses mutate all the time. This is why we need a new flu vaccine every year and why HIV drugs can eventually stop working. In fact, the flu pandemic of 1918 may have arisen from a bird flu through mutation.

It looks like something similar might have happened with the virus that causes AIDS, HIV. Apparently HIV arose from a virus found in chimpanzees called SIV.

But no one had ever seen SIV in wild populations of chimps. So we couldn't be sure where exactly HIV came from.

To solve this problem, a group of scientists set out to collect some chimpanzee poop and look for signs of the virus. What they found in southern Cameroon in Africa was poop that had SIV.

And not just any SIV but a new strain that was much more similar to HIV than were other strains. By looking at all of the viruses studied to date, scientists have concluded that this is the virus that jumped to humans and became HIV.

Most likely, someone ate SIV-infected chimpanzee meat so that the virus was now in people. The virus mutated and could now move from person to person.

The virus then made its way down the Sangha River to the Congo River to the city of Kinshasa. From there, the virus took off and spread across the world. We know all of this because of chimp poop.

We can definitely learn a lot from what people and animals leave behind. We catch criminals because of fingerprints, blood, hair, etc. And now we can learn so much about animals from their feces. Yucky but cool.

Answered by Dr. Barry Starr, Stanford University

Appendix – Punnett squares

> **I have brown eyes and my husband has blue eyes. We have three daughters: one has blue eyes, one green and one brown. How is this possible?**
>
> A curious adult from California

TO EXPLAIN WHY you can have brown-, green-, and blue-eyed kids, we need to go into a bit of detail about eye-color genetics. In the simplest models of eye color, there are two genes involved. (This model is too simple to explain a lot of things, but it will explain your situation.) Remember that for each gene, we inherit two copies, one from our mom and one from our dad.

For gene 1, there are two possibilities, brown or blue. The brown version of gene 1 is dominant over the blue one. What dominant means is if at least 1 of your two copies is brown, then you will have brown eyes. The way geneticists represent these two different versions of this eye-color gene are *B* for brown and *b* for blue (the capital letter is always the dominant, the lowercase, the recessive). So you will have brown eyes with either *BB* or *Bb* and blue eyes with *bb*.

For gene 2, there are two possibilities, green or blue. Green is dominant over blue and so *G* usually represents green and *b*, blue. Green eyes, then, can be *GG* or *Gb* while blue eyes are *bb*.

With me so far? One other thing you need to know is that brown is dominant over green—if you have a *B* version of gene 1 and a *G* version of gene 2, you will have brown eyes. Given all this, here are the possible gene combinations that can give you brown, green, or blue eyes:

BB bb	Brown
BB Gb	Brown
BB GG	Brown
Bb bb	Brown
Bb Gb	Brown
Bb GG	Brown
bb GG	Green
bb Gb	Green
bb bb	Blue

OK, what does this mean for you? Well, your husband can only be *bb bb* as he has blue eyes. Since you have brown eyes, you could have any of six different possibilities. But since you have brown-eyed, green-eyed and blue-eyed children, the most likely possibility for you is *Bb Gb*. You have brown eyes but are a carrier* for both blue and green eyes.

So how did I come up with your genotype (genetic makeup) as *Bb Gb*? As I'll try to show below, it is the only possibility where you could have green- or blue-eyed children. To help make this clearer, I'll introduce you to Punnett squares. Punnett squares are a relatively easy way to figure out if something is possible genetically and how likely it is.

The way a Punnett square works is you make a table. We'll do an easy one first with just gene 1 (the brown- or blue-eye gene). The first step is to put your two possible gene versions on the top like this:

	B	b

---

* A carrier is someone who can pass on a trait without having that trait - the trait is hidden. In our example, if you are *Bb*, that means you have brown eyes but could pass blue eyes to your children.

Since we are saying you are a brown-eyed carrier of blue eyes, you have a *B* (brown) and a *b* (blue) version of gene 1. The next step is to put your husband's gene versions on the side of the table like this:

	B	b
b		
b		

As you can see, in this example, I put your brown eyes on top (*Bb*) and your husband's blue eyes (*bb*) on the side. Remember, you and your husband only contribute one version of gene 1 each. You can give either a *B* or a *b* to your kids, not both. The Punnett square gives you all four possibilities of you and your husband's combinations.

The next step is to fill in each square with the letters from the top or side to figure out what is possible. For example, in the first square, since there is a *B* from you and a *b* from your husband, a *Bb* goes in like this:

	B	b
b	Bb	
b		

This represents a child like you who has brown eyes (*B*) but carries a blue-eye version of gene 1 (*b*). You then fill in the rest of the square like this:

	B	b
b	Bb	bb
b	Bb	bb

From this, you can figure out that you have a 50-50 shot of having either blue-eyed or brown-eyed kids. In the table, there are two *Bb* and two *bb* squares meaning you have a 2 in 4 chance of brown-eyed kids (*Bb*) and a 2 in 4 chance of blue-eyed kids (*bb*). Note that all your brown-eyed kids will be carriers for the blue-eyed version of the gene, *b*.

Now, to add the green gene, it gets more complicated. Each of the two genes is independent of each other so you need to figure out all of the possibilities you could have. Your husband is easy as all his possibilities will be *bb*. For you, it is a little trickier.

Remember, you are most likely *Bb Gb*. So what should we put on the top of the square for you? You figure this out by starting with *B*. If one of your children gets *B*, then they can get either *G* or *b* from gene 2. The same is true for *b*. So the possibilities are *BG*, *Bb*, *bG*, and *bb*. The Punnett square for you and your husband would look like this:

	BG	Bb	bG	bb
bb				
bb				
bb				
bb				

You do the same thing as before and combine the boxes. The first box would be *Bb Gb*, a brown-eyed carrier of green and blue eyes. If we fill in all of the possibilities, we get:

	BG	Bb	bG	bb
bb	**Bb Gb**	**Bb bb**	bb Gb	bb bb
bb	**Bb Gb**	**Bb bb**	bb Gb	bb bb
bb	**Bb Gb**	**Bb bb**	bb Gb	bb bb
bb	**Bb Gb**	**Bb bb**	bb Gb	bb bb

From this, we can figure that you and your husband have a 50% chance of a brown-eyed child (anything with a *B*), a 25% chance of a green-eyed child (anything with a *G* and no *B*), and a 25% chance of a blue-eyed child (anything with just *b*).

So as you can see, if you really are a *Bb Gb*, then this genetic model explains your kids perfectly. You can try out other possible brown-eye genetic combinations and see whether or not you can get green- or blue-eyed children.

Answered by Dr. Barry Starr, Stanford University

Glossary

__ACGT__ is an acronym for the four types of bases found in a DNA molecule: adenine (A), cytosine (C), guanine (G), and thymine (T). A DNA molecule consists of two strands wound around each other, with each strand held together by bonds between the bases. Adenine pairs with thymine, and cytosine pairs with guanine. The sequence of bases in a portion of a DNA molecule, called a gene, carries the instructions needed to assemble a protein.[1]

An __allele__ is one of two or more versions of a gene. An individual inherits two alleles for each gene, one from each parent. If the two alleles are the same, the individual is homozygous for that gene. If the alleles are different, the individual is heterozygous. Though the term "allele" was originally used to describe variation among genes, it now also refers to variation among non-coding DNA sequence.[1]

__Autosomal dominance__ is a pattern of inheritance characteristic of some genetic diseases. "Autosomal" means that the gene in question is located on one of the numbered, or non-sex, chromosomes. "Dominant" means that a single copy of the disease-associated mutation is enough to cause the disease. This is in contrast to a recessive disorder, where two copies of the mutation are needed to cause the disease. [1]

An __autosome__ is any of the numbered chromosomes, as opposed to the sex chromosomes. Humans have 22 pairs of autosomes and one pair of sex chromosomes (the X and Y). Autosomes are numbered roughly in relation to their sizes. That is, chromosome 1 has approximately 2,800 genes, while chromosome 22 has approximately 750 genes.[1]

---

1   Definition courtesy of the National Human Genome Research Institute.

A **base pair** is two chemical bases bonded to one another forming a "rung of the DNA ladder." The DNA molecule consists of two strands that wind around each other like a twisted ladder. Each strand has a backbone made of alternating sugar (deoxyribose) and phosphate groups. Attached to each sugar is one of four bases–adenine (A), cytosine (C), guanine (G), or thymine (T). The two strands are held together by hydrogen bonds between the bases, with adenine forming a base pair with thymine, and cytosine forming a base pair with guanine.[1]

A **carrier** is an individual who carries and is capable of passing on a genetic mutation associated with a disease and may or may not display disease symptoms. Carriers are associated with diseases inherited as recessive traits. In order to have the disease, an individual must have inherited mutated alleles from both parents. An individual having one normal allele and one mutated allele does not have the disease. Two carriers may produce children with the disease.[1]

A **cell** is the building block of all living things. Cells in plants and animals have different compartments. One of these compartments is called the nucleus. The nucleus is where DNA, packaged into chromosomes, is found.

A **chromosome** is an organized package of DNA found in the nucleus of the cell. Different organisms have different numbers of chromosomes. Humans have 23 pairs of chromosomes—22 pairs of numbered chromosomes, called autosomes, and one pair of sex chromosomes, X and Y. Each parent contributes one chromosome to each pair so that offspring get half of their chromosomes from their mother and half from their father.[1]

DNA is the chemical name for the molecule that carries genetic instructions in all living things. The DNA molecule consists of two strands that wind around one another to form a shape known as a double helix. Each strand has a backbone made of alternating sugar (deoxyribose) and phosphate groups. Attached to each sugar is one of four bases–adenine (A), cytosine (C), guanine (G), and thymine (T). The two strands are held together by bonds between the bases; adenine bonds with thymine, and cytosine bonds with guanine. The sequence of the bases along the backbones serves as instructions for assembling protein and RNA molecules.[1]

DNA replication is the process by which a molecule of DNA is duplicated. When a cell divides, it must first duplicate its genome so that each daughter cell winds up with a complete set of chromosomes.[1]

Dominant refers to the relationship between two versions of a gene. Individuals receive two versions of each gene, known as alleles, from each parent. If the alleles of a gene are different, one allele will be expressed; it is the dominant gene. The effect of the other allele, called recessive, is masked.[1]

Epigenetics is an emerging field of science that studies heritable changes caused by the activation and deactivation of genes without any change in the underlying DNA sequence of the organism. The word epigenetics is of Greek origin and literally means over and above (epi) the genome.[1]

Fraternal twins are also dizygotic twins. They result from the fertilization of two separate eggs during the same pregnancy. Fraternal twins may be of the same or different sexes. They share half of their genes just like any other siblings. In contrast, twins that result from the fertilization of a single egg that then splits in two are called monozygotic, or identical, twins. Identical twins share all of their genes and are always the same sex.[1]

The **gene** is the basic physical unit of inheritance. Genes are passed from parents to offspring and contain the information needed to specify traits. Genes are arranged, one after another, on structures called chromosomes. A chromosome contains a single, long DNA molecule, only a portion of which corresponds to a single gene. Humans have approximately 23,000 genes arranged on their chromosomes.[1]

Gene–environment interaction is an influence on the expression of a trait that results from the interplay between genes and the environment. Some traits are strongly influenced by genes, while other traits are strongly influenced by the environment. Most traits, however, are influenced by one or more genes interacting in complex ways with the environment.[1]

Gene expression is the process by which the information encoded in a gene is used to direct the assembly of a protein molecule. The cell reads the sequence of the gene in groups of three bases. Each group of three bases (codon) corresponds to one of 20 different amino acids used to build the protein.[1]

Genetic code. The instructions in a gene that tell the cell how to make a specific protein. A, C, G, and T are the "letters" of the DNA code; they stand for the chemicals adenine (A), cytosine (C), guanine (G), and thymine (T), respectively, that make up the nucleotide bases of DNA. Each gene's code combines the four chemicals in various ways to spell out three-letter "words" that specify which amino acid is needed at every step in making a protein.[1]

Genetic engineering is the process of using recombinant DNA (rDNA) technology to alter the genetic makeup of an organism. Traditionally, humans have manipulated genomes indirectly by controlling breeding and selecting offspring with desired traits. Genetic engineering involves the direct manipulation of one or more genes. Most often, a gene from another species is added to an organism's genome to give it a desired phenotype.[1]

The **genome** is the entire set of genetic instructions found in a cell. In humans, the genome consists of 23 pairs of chromosomes, found in the nucleus, as well as a small chromosome found in the cells' mitochondria. These chromosomes, taken together, contain approximately 3.1 billion bases of DNA sequence.[1]

Identical twins are also known as monozygotic twins. They result from the fertilization of a single egg that splits in two. Identical twins share all of their genes and are always of the same sex. In contrast, fraternal, or dizygotic, twins result from the fertilization of two separate eggs during the same pregnancy. They share half of their genes, just like any other siblings. Fraternal twins can be of the same or different sexes.[1]

An **inherited** trait is one that is genetically determined. Inherited traits are passed from parent to offspring according to the rules of Mendelian genetics. Most traits are not strictly determined by genes, but rather are influenced by both genes and environment.[1]

Meiosis is the formation of egg and sperm cells. In sexually reproducing organisms, body cells are diploid, meaning they contain two sets of chromosomes (one set from each parent). To maintain this state, the egg and sperm that unite during fertilization must be haploid, meaning they each contain a single set of chromosomes. During meiosis, diploid cells undergo DNA replication, followed by two rounds of cell division, producing four haploid sex cells.[1]

Melanin is a chemical that gives color to skin, hair, and eyes, and helps protect them from damage from UV light. There are two kinds of melanin: brown/black eumelanin and red/yellow pheomelanin.

Melanocytes are special cells that make the pigment melanin.

DNA **methylation** is the addition of a chemical group, called a methyl group, to the DNA molecule. It affects which genes are turned off. DNA methylation patterns are inherited but are not permanent—they can change in response to the environment.

Mitosis is a cellular process that replicates chromosomes and produces two identical nuclei in preparation for cell division. Generally, mitosis is immediately followed by the equal division of the cell nuclei and other cell contents into two daughter cells.[1]

A **modifier gene** is a gene that affects the expression of other genes.

A **mutation** is a change in a DNA sequence. Mutations can result from DNA copying mistakes made during cell division, exposure to ionizing radiation, exposure to chemicals called mutagens, or infection by viruses. Germ line mutations occur in the eggs and sperm and can be passed on to offspring, while somatic mutations occur in body cells and are not passed on.[1]

Penetrance refers to the fraction of people with a gene version for a trait who actually show that trait. Penetrance can be complete or incomplete (variable). For example, a version of the gene *FGFR3* that causes dwarfism shows complete penetrance—everybody who inherits one copy of that gene version is a dwarf. Cleft chin is an example of variable, or incomplete, penetrance—only some of the people who inherit a cleft-chin gene have a chin with a cleft.

A **phenotype** is an individual's observable traits, such as height, eye color, and blood type. The genetic contribution to the phenotype is called the genotype. Some traits are largely determined by the genotype, while other traits are largely determined by environmental factors.[1]

Proteins are an important class of molecules found in all living cells. A protein is composed of one or more long chains of amino acids, the sequence of which corresponds to the DNA sequence of the gene that encodes it. Proteins play a variety of roles in the cell, including structural (cytoskeleton), mechanical (muscle), biochemical (enzymes), and cell signaling (hormones). Proteins are also an essential part of diet.[1]

A **Punnett square** is a chart that lists all possible combinations of gene versions (for one or more genes) a mom and a dad can pass on to their kids. It can also reveal the chances that a child will end up with a particular trait if the genes involved are understood well enough. For details on how to build and interpret a Punnett square, see the Appendix.

Recessive is a quality found in the relationship between two versions of a gene. Individuals receive one version of a gene, called an allele, from each parent. If the alleles are different, the dominant allele will be expressed, while the effect of the other allele, called recessive, is masked. In the case of a recessive genetic disorder, an individual must inherit two copies of the mutated allele in order for the disease to be present.[1]

The amniotic **sac** is the part of the womb that holds the developing fetus and the amniotic fluid that surrounds the fetus.

A **sex chromosome** is a type of chromosome that participates in sex determination. Humans and most other mammals have two sex chromosomes, the X and the Y. Females have two X chromosomes in their cells, while males have both X and a Y chromosomes in their cells. Egg cells all contain an X chromosome, while sperm cells contain an X or Y chromosome. This arrangement means that it is the male that determines the sex of the offspring when fertilization occurs.[1]

A **somatic cell** is any cell of the body except sperm and egg cells. Somatic cells are diploid, meaning that they contain two sets of chromosomes, one inherited from each parent. Mutations in somatic cells can affect the individual, but they are not passed on to offspring.[1]

The **stroma** of the iris is the front layer of the iris. It holds the pigment (melanin) that gives the eye its color.

A **trait** is a specific characteristic of an organism. Traits can be determined by genes or the environment, or more commonly by interactions between them. The genetic contribution to a trait is called the genotype. The outward expression of the genotype is called the phenotype.[1]

A **zygote** is a cell that results from the fusion of a sperm cell with an egg.

Bibliography

page 2 temperature-dependence of sex determination in turtles

Ewert MA and Nelson CE. Sex Determination in Turtles: Diverse Patterns and Some Possible Adaptive Values. *Copeia* (1991) 1:50-69.

page 2 sex determination in clownfish

Buston P. Social hierarchies: size and growth modification in clownfish. *Nature* (2003) 424:145-146.

page 9 *SCN9A* mutation

Cox JJ, Reimann F, Nicholas AK, Thornton G, Roberts E, Springell K, Karbani G, Jafri H, Mannan J, Raashid Y, Al-Gazali L, Hamamy H, Valente EM, Gorman S, Williams R, McHale DP, Wood JN, Gribble FM, and Woods CG. An *SCN9A* channelopathy causes congenital inability to experience pain. *Nature* (2006) 444:894-898.

page 10 doogie mouse

Tang YP, Shimizu E, Dube GR, Rampon C, Kerchner GA, Zhuo M, Liu G, and Tsien JZ. Genetic enhancement of learning and memory in mice. *Nature* (1999) 401:63-69.

page 11 myostatin mutation

Schuelke M, Wagner KR, Stolz LE, Hubner C, Riebel T, Komen W, Braun T, Tobin JF, and Lee SJ. Myostatin mutation associated with gross muscle hypertrophy in a child. *N Engl J Med* (2004) 350:2682-2688.

page 13 two-headed snake attacking other head

Mayell H. Life Is Confusing For Two-Headed Snakes. In *National Geographic News*, March 22, 2002.

page 13 twin sac and placenta configuration

Embleton ND and Pillalamarri KVS. Whistle blowing in clinical diagnosis. *Archives of disease in childhood - Education & practice edition* (2007) 92:ep70-ep75.

page 13 frequency of conjoined twins

Kaufman MH. The embryology of conjoined twins. *Childs Nerv Syst* (2004) 20:508-525.

page 14 attachment sites of conjoined twins

Baron BW, Shermeta DW, Ismail MA, Ben-Ami T, Yousefzadeh D, Carlson N, Amarose AP, and Esterly JR. Unique anomalies in cephalothoracopagus janiceps conjoined twins with implications for multiple mechanisms in the abnormal embryogenesis. *Teratology* (1990) 41:9-22.

page 15 *GLI3* and polydactyly

Online Mendelian Inheritance in Man, OMIM™. Johns Hopkins University, Baltimore, MD. MIM Number: 165240: 8/04/2011. http://www.ncbi.nlm.nih.gov/omim/

page 15 *GLI3* and *HAND2*

Hill P, Wang B, and Ruther U. The molecular basis of Pallister-Hall associated polydactyly. *Hum Mol Genet* (2007) 16:2089-2096.

page 19 rice that makes vitamin A

Paine JA, Shipton CA, Chaggar S, Howells RM, Kennedy MJ, Vernon G, Wright SY, Hinchliffe E, Adams JL, Silverstone AL, and Drake R. Improving the nutritional value of Golden Rice through increased pro-vitamin A content. *Nat Biotechnol* (2005) 23:482-487.

Ye X, Al-Babili S, Kloti A, Zhang J, Lucca P, Beyer P, and Potrykus I. Engineering the provitamin A (beta-carotene) biosynthetic pathway into (carotenoid-free) rice endosperm. *Science* (2000) 287:303-305.

page 20 cacao genome

Argout X, Salse J, Aury JM, Guiltinan MJ, Droc G, Gouzy J, Allegre M, Chaparro C, Legavre T, Maximova SN, Abrouk M, Murat F, Fouet O, Poulain J, Ruiz M, Roguet Y, Rodier-Goud M, Barbosa-Neto JF, Sabot F, Kudrna D, Ammiraju JS, Schuster SC, Carlson JE, Sallet E, Schiex T, Dievart A, Kramer M, Gelley L, Shi Z, Berard A, Viot C, Boccara M, Risterucci AM, Guignon V, Sabau X, Axtell MJ, Ma Z, Zhang Y, Brown S, Bourge M, Golser W, Song X, Clement D, Rivallan R, Tahi M, Akaza JM, Pitollat B, Gramacho K, D'Hont A, Brunel D, Infante D, Kebe I, Costet P, Wing R, McCombie WR, Guiderdoni E, Quetier F, Panaud O, Wincker P, Bocs S, and Lanaud C. The genome of Theobroma cacao. *Nat Genet* (2011) 43:101-108.

Yang JL. With DNA of chocolate nearly decoded by scientists, could sweeter treats await? In *The Washington Post*, September 15, 2010.

page 28 frequency of Europeans' eye color

Sturm RA, Duffy DL, Zhao ZZ, Leite FP, Stark MS, Hayward NK, Martin NG, and Montgomery GW. A single SNP in an evolutionary conserved region within intron 86 of the *HERC2* gene determines human blue-brown eye color. *Am J Hum Genet* (2008) 82:424-431.

https://www.23andme.com/health/Eye-Color/

page 35 Waardenburg syndrome

Karaman A and Aliagaoglu C. Waardenburg syndrome type 1. *Dermatology online journal* (2006) 12:21.

page 36 David Bowie's eye color

Buckley D. *Strange Fascination. David Bowie: The Definitive Story*, Virgin Books Ltd, London, 2002.

page 38 eye-color changes

Bito LZ, Matheny A, Cruickshanks KJ, Nondahl DM, and Carino OB. Eye color changes past early childhood. The Louisville Twin Study. *Arch Ophthalmol* (1997) 115:659-663.

page 43 melanin packaging

Rees JL. Genetics of hair and skin color. *Annu Rev Genet* (2003) 37:67-90.

page 44 *KITLG* and pigmentation in fish and humans

Miller CT, Beleza S, Pollen AA, Schluter D, Kittles RA, Shriver MD, and Kingsley DM. cis-Regulatory changes in *Kit ligand* expression and parallel evolution of pigmentation in sticklebacks and humans. *Cell* (2007) 131:1179-1189.

page 44 *SLC24A5* and pigmentation in fish and humans

Lamason RL, Mohideen MA, Mest JR, Wong AC, Norton HL, Aros MC, Jurynec MJ, Mao X, Humphreville VR, Humbert JE, Sinha S, Moore JL, Jagadeeswaran P, Zhao W, Ning G, Makalowska I, McKeigue PM, O'Donnell D, Kittles R, Parra EJ, Mangini NJ, Grunwald DJ, Shriver MD, Canfield VA, and Cheng KC. *SLC24A5*, a putative cation exchanger, affects pigmentation in zebrafish and humans. *Science* (2005) 310:1782-1786.

page 44 light skin color in Europeans

Makova K and Norton H. Worldwide polymorphism at the MC1R locus and normal pigmentation variation in humans. *Peptides* (2005) 26:1901-1908.

Raimondi S, Sera F, Gandini S, Iodice S, Caini S, Maisonneuve P, and Fargnoli MC. *MC1R* variants, melanoma and red hair color phenotype: a meta-analysis. *Int J Cancer* (2008) 122:2753-2760.

page 48 melanin in skin cells

Morelli JG. (2011) Hyperpigmented Lesions, In *Nelson Textbook of Pediatrics* (Kliegman RM, Stanton BMD, St. Geme J, Schor N, and Behrman RE, Eds.) 19th ed., Elsevier/Saunders, Philadelphia, PA.

page 48 *MC1R* is the major freckle gene

Bastiaens M, ter Huurne J, Gruis N, Bergman W, Westendorp R, Vermeer BJ, and Bouwes Bavinck JN. The melanocortin-1-receptor gene is the major freckle gene. *Hum Mol Genet* (2001) 10:1701-1708.

page 49 freckle gene on chromosome 4

Zhang XJ, He PP, Liang YH, Yang S, Yuan WT, Xu SJ, and Huang W. A gene for freckles maps to chromosome 4q32-q34. *J Invest Dermatol* (2004) 122:286-290.

page 49 freckle genes in Europeans

Sulem P, Gudbjartsson DF, Stacey SN, Helgason A, Rafnar T, Magnusson KP, Manolescu A, Karason A, Palsson A, Thorleifsson G, Jakobsdottir M, Steinberg S, Palsson S, Jonasson F, Sigurgeirsson B, Thorisdottir K, Ragnarsson R, Benediktsdottir KR, Aben KK, Kiemeney LA, Olafsson JH, Gulcher J, Kong A, Thorsteinsdottir U, and Stefansson K. Genetic determinants of hair, eye and skin pigmentation in Europeans. *Nat Genet* (2007) 39:1443-1452.

page 50 *Mc1r* variation in mice

Hoekstra HE, Hirschmann RJ, Bundey RA, Insel PA, and Crossland JP. A single amino acid mutation contributes to adaptive beach mouse color pattern. *Science* (2006) 313:101-104.

page 50 *Mc1r* variation in the mammoth

Rompler H, Rohland N, Lalueza-Fox C, Willerslev E, Kuznetsova T, Rabeder G, Bertranpetit J, Schoneberg T, and Hofreiter M. Nuclear gene indicates coat-color polymorphism in mammoths. *Science* (2006) 313:62.

page 52-53 hair-color genes in Europeans

Sulem P, Gudbjartsson DF, Stacey SN, Helgason A, Rafnar T, Jakobsdottir M, Steinberg S, Gudjonsson SA, Palsson A, Thorleifsson G, Palsson S, Sigurgeirsson B, Thorisdottir K, Ragnarsson R, Benediktsdottir KR, Aben KK, Vermeulen SH, Goldstein AM, Tucker MA, Kiemeney LA, Olafsson JH, Gulcher J, Kong A, Thorsteinsdottir U, and Stefansson K. Two newly identified genetic determinants of pigmentation in Europeans. *Nat Genet* (2008) 40:835-837.

Sulem P, Gudbjartsson DF, Stacey SN, Helgason A, Rafnar T, Magnusson KP, Manolescu A, Karason A, Palsson A, Thorleifsson G, Jakobsdottir M, Steinberg S, Palsson S, Jonasson F, Sigurgeirsson B, Thorisdottir K, Ragnarsson R, Benediktsdottir KR, Aben KK, Kiemeney LA, Olafsson JH, Gulcher J, Kong A, Thorsteinsdottir U, and Stefansson K. Genetic determinants of hair, eye and skin pigmentation in Europeans. *Nat Genet* (2007) 39:1443-1452.

page 60 *AR* gene and baldness

Hillmer AM, Hanneken S, Ritzmann S, Becker T, Freudenberg J, Brockschmidt FF, Flaquer A, Freudenberg-Hua Y, Jamra RA, Metzen C, Heyn U, Schweiger N, Betz RC, Blaumeiser B, Hampe J, Schreiber S, Schulze TG, Hennies HC, Schumacher J, Propping P, Ruzicka T, Cichon S, Wienker TF, Kruse R, and Nothen MM. Genetic variation in the human androgen receptor gene is the major determinant of common early-onset androgenetic alopecia. *Am J Hum Genet* (2005) 77:140-148.

page 61 baldness and chromosome 20

Hillmer AM, Brockschmidt FF, Hanneken S, Eigelshoven S, Steffens M, Flaquer A, Herms S, Becker T, Kortum AK, Nyholt DR, Zhao ZZ, Montgomery GW, Martin NG, Muhleisen TW, Alblas MA, Moebus S, Jockel KH, Brocker-Preuss M, Erbel R, Reinartz R, Betz RC, Cichon S, Propping P, Baur MP, Wienker TF, Kruse R, and Nothen MM. Susceptibility variants for male-pattern baldness on chromosome 20p11. *Nat Genet* (2008) 40:1279-1281.

Richards JB, Yuan X, Geller F, Waterworth D, Bataille V, Glass D, Song K, Waeber G, Vollenweider P, Aben KK, Kiemeney LA, Walters B, Soranzo N, Thorsteinsdottir U, Kong A, Rafnar T, Deloukas P, Sulem P, Stefansson H, Stefansson K, Spector TD, and Mooser V. Male-pattern baldness susceptibility locus at 20p11. *Nat Genet* (2008) 40:1282-1284.

page 62 *FGFR3* and dwarfism

Shiang R, Thompson LM, Zhu YZ, Church DM, Fielder TJ, Bocian M, Winokur ST, and Wasmuth JJ. Mutations in the transmembrane domain of FGFR3 cause the most common genetic form of dwarfism, achondroplasia. *Cell* (1994) 78:335-342.

page 62 two copies of dwarfism variants of *FGFR3* are lethal

Rousseau F, Bonaventure J, Legeai-Mallet L, Pelet A, Rozet JM, Maroteaux P, Le Merrer M, and Munnich A. Mutations in the gene encoding fibroblast growth factor receptor-3 in achondroplasia. *Nature* (1994) 371:252-254.

page 63 *FGFR3* mutations that appear out of nowhere

Richette P, Bardin T, and Stheneur C. Achondroplasia: from genotype to phenotype. *Joint Bone Spine* (2008) 75:125-130.

page 66 frequency of chimeras

Ainsworth C. The stranger within. *New Scientist* (2003) 180:34.

page 79 13 places for DNA typing

Butler JM. *Fundamentals of Forensic DNA Typing*, Academic Press, 2009.

page 80 crime suspect with bone marrow transplant

Aldhous P. Bone marrow donors risk DNA identity mix-up. *New Scientist* (2005) 188:11.

page 85 crime suspects are twins

Willing R. Twin suspects spark unique DNA test. In *USA TODAY*, September 1, 2004.

page 86 grass in dinosaur poop

Prasad V, Stromberg CA, Alimohammadian H, and Sahni A. Dinosaur coprolites and the early evolution of grasses and grazers. *Science* (2005) 310:1177-1180.

page 87 counting pandas using poop

Zhan X, Li M, Zhang Z, Goossens B, Chen Y, Wang H, Bruford MW, and Wei F. Molecular censusing doubles giant panda population estimate in a key nature reserve. *Curr Biol* (2006) 16:R451-452.

page 89 viruses in chimpanzee poop

Keele BF, Van Heuverswyn F, Li Y, Bailes E, Takehisa J, Santiago ML, Bibollet-Ruche F, Chen Y, Wain LV, Liegeois F, Loul S, Ngole EM, Bienvenue Y, Delaporte E, Brookfield JF, Sharp PM, Shaw GM, Peeters M, and Hahn BH. Chimpanzee reservoirs of pandemic and nonpandemic HIV-1. *Science* (2006) 313:523-526.

Index